Penguin Books
Bird of Life, Bird of Death

Ornithologist and journalist Jonathan Evan Maslow writes on
natural history, politics, sports and travel. His work has appeared in
GEO, *Atlantic Monthly*, *The New York Times* and the *Boston
Globe*, among other periodicals. He makes his home in Cape May
Country, New Jersey. His first book, *The Owl Papers*, is published
in Penguin.

Jonathan Evan Maslow

Bird of Life, Bird of Death

A Naturalist's Journey Through a Land
of Political Turmoil

Penguin Books

Penguin Books Ltd, Harmondsworth, Middlesex, England
Viking Penguin Inc., 40 West 23rd Street, New York, New York 10010, U.S.A.
Penguin Books Australia Ltd, Ringwood, Victoria, Australia
Penguin Books Canada Ltd, 2801 John Street, Markham, Ontario, L3R 1B4
Penguin Books (N.Z.) Ltd, 182–190 Wairau Road, Auckland 10, New Zealand

First published in the U.S.A. by Simon and Schuster,
a division of Simon & Schuster, Inc. 1986
First published in Great Britain by Viking 1986
Published in Penguin Books 1987

Made and printed in Great Britain by
Richard Clay Ltd, Bungay, Suffolk
Typeset in Monophoto Sabon

For My Parents
and
for David Maxey
In memoriam

Contents

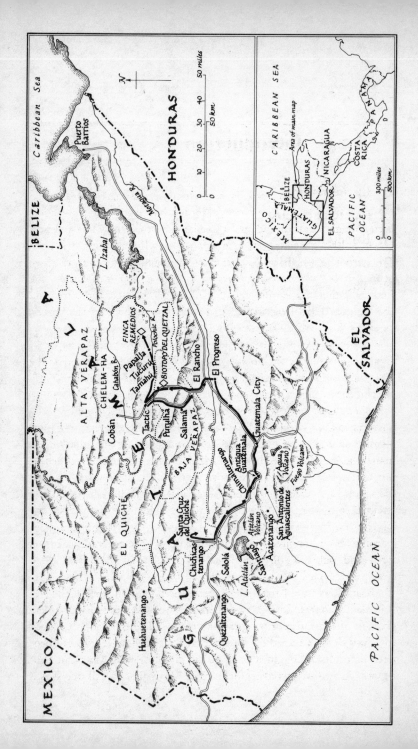

Introduction

This book is not, strictly speaking, a bird book, though it was written by a fanatical birder. Nor is it a true travelogue, even though I am an addicted traveller. Least of all is it a political history, although, like most men, I fancy myself something of a political expert. But this book does borrow shamelessly from all three genres. It is the impressionistic account of a trip I took into the Central American cordillera of Guatemala to see the rare and endangered Resplendent Quetzal, to learn something of that bird's extensive lore, and to investigate its impending extinction.

What motivated the author? An obsessive curiosity about nature, I suppose, and maybe the dream of seeing a bird long renowned as the most handsome and legendary in all the Americas. I had recently concluded an unpaid, year-long vacation studying owls for a book called *The Owl Papers*, and emerged from that experience with some answers to a question I had had about my lifelong interest in natural history: namely, why I never fully believed anything about anything, man or beast, until I saw, heard, smelled, or tasted – in short, witnessed – it for myself. The answers: more than a dash of egotism, an easy seduction by nature's sensuality, some anti-rationalist obstinacy, the positive enjoyment of difficult endeavours, and the tawdry taste for adventures, preferably outdoors. Above all, the intoxication of learning about life directly through the gut, of rediscovering nature as fresh and palpable on every field-trip. You may recognize these symptoms.

Around this time I read Alexander Skutch's life history of the Resplendent Quetzal, dating from the 1940s but still the most complete account of *Pharomachrus mocinno* we have. There I was introduced to a bird of such incredible beauty that for two hundred years European naturalists thought it must be the fabrication of American aborigines. A bird so sacred to the ancient Maya that to kill one was a capital crime. A bird so closely associated with its lofty home in the Central American cloud forests that as the forests vanished, so too did the bird. A bird that has been a symbol of liberty for several *thousand* years – not the shrill, defiant liberty of the eagle, but the serene and innocent liberty of the child at play. That this unique bird is the national symbol of Guatemala, a nation often cited as one of the most repressive on earth, was an irony that almost alone made a trip worthwhile. The engine of my curiosity was instantly engaged.

Anyone who has followed the tragic news coming out of Central America in recent years will realize that deciding to travel there is not something one does with a light heart or an easy mind. But I had visited Latin America on numerous occasions, and lived there twice for lengthy periods of time. As the turmoil and violence rose in Central America, I found myself drawn there to write about politics, some-times with a magazine commission, sometimes without. Thus, at the outset, I at least had some concrete idea of what I was getting myself into. It was then that I met the photographer Michael Kienitz, a chance acquaintance who would become a fast friend. I first spotted him on the Moskito Coast near the Nicaraguan border with Hon-duras, where he was photographing the war between the Sandinistas and the Contras. It was just after a vicious fire fight that left seventy young men dead and an entire village in flames. Mickey couldn't take his camera off the Black Vultures feeding on the corpses, dismembering the bodies in rough shifts, like factory workers. Every frame was filled with images of the horror of Central America. It was an introduction to the unnatural history of that region one was not likely to forget.

Unlike me, Mickey was not one to light out after magical birds, no matter how alluring. Yet as a fully uninsured member of the fraternity of the road, he shared my vocation as a witness. When I suggested we go into the Guatemalan highlands together after the Quetzal, he agreed immediately. This made the trip possible for me; I would never have contemplated it alone. I would certainly have died of frustration, loneliness, fear, or possibly just from a stray bullet. If I took advantage of my new partner by enlisting him in my bizarre mission, I am sorry for it, but only to a degree.

At the time we planned our trip, the country had been under continu-ous military dictatorship since 1954. Its latest incarnation was the spooky evangelical General Efraín Ríos Montt. In the tradition of

Latin American tyrants, Montt had come to power in a putsch, promising democratic elections, then instituted a reign of terror in the countryside. He announced on national television that he felt inspired by God to continue in office until all subversion and rebellion in the fatherland were crushed. That could be a long proposition. While the general rehearsed his immortality, the mounds of the dead heaped up. Casualty estimates ranged from the low thousands to the tens of thousands. Most of the victims, according to human-rights monitors, were Indian peasants in remote rural regions of the country. Several provinces, including those in the central highlands where the Quetzal resides, had been closed to foreigners. As a result, we were forced to delay our departure, missing the tropical dry season when Quetzals might be readily sighted at their nesting places. In addition, we could obtain only thirty-day tourist visas, the device authoritarian regimes favour for controlling snoopy foreigners. These factors added immensely to our difficulties. Practical information about reaching a Quetzal habitat was practically nil. Time and free movement, which are basic necessities for watching wildlife, seemed out of the question. Nevertheless, when Mickey arrived on my doorstep with two shiny green Gore-Tex rain suits, and the news that foreigners were being allowed to travel in the Guatemalan highlands again, I knew it was time to go. We set out for Guatemala in July 1983.

Although Central America is a region that tends to obliterate the discreet categories we usually use to classify our experiences – natural science blends into the supernatural, natural history merges with legend, historical reality couples with fiction – I've tried in writing this book to keep my feet on the ground at all times. The people sketched here are real people (and may the gods keep them safe from the dangers of exposure). The incidents happened. It is not true, as is often said, that life is held cheap in Central America, but it is true that life is briefer there, and so takes place faster. I have slowed things down in places, to keep the story comprehensible to the reader, and in other places, I have digressed, to keep attention focused on the bird. I may be justly accused of trying to promote the premise that what happens to the illustrious Quetzal of Central America is inextricably bound with human destiny there. If anthropomorphism is the attempt to read human motives into animal behaviour, then I have headed in the opposite direction at sixty miles an hour: towards a bird whose fate might illuminate some of humanity's brightest and darkest moments. In writing about the Quetzal, I have attempted a kind of essay in political ornithology – a field that does not quite exist, at least yet.

For their help in the making of this book, I would like to acknowledge and thank my wife, Sarah; Kathy Robbins; Bob Bender; Stephanie Franklin; and Jeannie de Ford.

I

Guatemala City

I

It was the rainy season in Guatemala when we set off in search of the rare and beautiful Quetzal, and the general's public relations flak was reluctant to give us safe conducts for travel to the bird's habitat in the highlands. For some minutes he sat behind his carved desk in the Presidential Palace in Guatemala City, casting his eyes over our letter of introduction like a vulture studying a carcass, trying to decide if it's sufficiently rotten to eat. Pudgy, sallow-faced, funereally slick in a dark pencil-stripe suit, he had already kept my partner Mickey and me waiting an hour in his anteroom, so waiting a few minutes more while he decided how to get rid of us did not matter. At length, he lifted his heavy-lidded eyes and said, 'Frankly, where you propose to go to observe our national bird, the Quetzal, is a zone of conflict. I don't recommend going there – but I don't recommend not going there either. Guatemala is a free country; you can go where you like. Unfortunately, I don't see how I can help you. If you encounter the army during your expedition, you won't need any credentials. And if you meet the guerrillas? Well, a letter from the palace won't do you any good.'

So the verbal duel was engaged. Leaving aside for the moment the more than 10,000 abductions, assassinations, and disappearances attributed to the Guatemalan military and the paramilitary death squads by human-rights groups in 1983, I argued that (1) the optimal hours for bird-watching are just after dawn, often necessitating movement to an observation site during hours of darkness, when much of Central America becomes a free-fire zone; (2) we were carrying fieldglasses and a birder's spotting scope mounted on a gun stock, both easily mistaken for military equipment, and (3) while credentials might be less important for me, Mickey's photographic gear might arouse suspicion among those in the countryside unused to dealing with foreigners. A letter might avoid a situation that could, I reminded him pointedly, reflect poorly on the freedom-loving security forces of Guatemala.

Señor Escobar chewed this over with a condescending nod of touché, then ordered his secretary in to dial the telephone for him: he proposed to contact a certain professor at the University of San Carlos. The professor was overseer of conservation programmes at the University's Biotopo del Quetzal, the wildlife sanctuary in Baja Verapaz province devoted to preserving the Quetzal's cloud forest habitat. If the professor would grant us a permit to do our Quetzal watching in the Biotopo, we could go there, and the matter would be closed.

Parry and thrust: what weight would a permit from some university professor carry out on the murderous roads of Central America? Ornithologists we had consulted before leaving for Guatemala were unsure if the Biotopo was even still functioning; that part of Guatemala had been officially declared off-limits to foreigners for more than a year. Everyone had heard the rumours of fierce violence in the area. They had recommended we drop the whole idea of finding Quetzals in Guatemala, or at the very least go armed with IDs, credentials, letters of introduction, one from the palace, another from army headquarters, even from local commanders and mayors. The idea was to deploy the appropriate document at the right time, respecting the chain of command yet letting the particular party hassling us know that his immediate superiors in the hierarchy had stamped approval on our heads. It sounded like overkill, perhaps, but better that than the real thing.

While the secretary undertook the heavy manual labour of dialling the telephone, Señor Escobar sat smiling contentedly, his hands folded neatly round his paunch, in which position he looked like Al Capone posing as the Buddha.

Looking round the room in the interval, Mickey said, in markedly loud and clear English, 'Notice something funny about this office? There's no picture of the general on the wall. I guess this guy doesn't expect his boss to stay in power very long.'

Escobar squirmed and lost his beatific smile, as the secretary reported that the professor wasn't in. Assuring us he'd get back to us by the end of the work week next day, he ceremoniously showed us the door. So much for help from the national authorities.

Down the portico facing inwards to the courtyard a dark knot of newsmen hunched outside the men's room. I asked whom they were waiting for, and one said they were waiting for General Ríos Montt himself, who was inside the gents'. 'Does he spend a lot of time in there?' I asked. 'Lately, quite a lot,' the reporter chuckled.

Only a few days had passed since the general's enemies – other generals – had attempted a *coup d'état* against him. It fizzled when some military units remained loyal to the general's regime, but the palace was still shaky, and word was around that the general's days in power were numbered. The general talked regularly to God, and relayed God's word to his nation in windy television sermons – he was an ordained deacon in an evangelical Protestant sect founded by ex-hippies from California. In March of 1982, amidst civil disturbances protesting widespread voting fraud in presidential elections, the other generals had put their troops in the streets and occupied the palace. They called in Ríos Montt as an afterthought. He was retired at the time, a harmless old soldier deeply engrossed in his zealous evangelism. The other generals thought he would be a good front man for Guatemala's international image. They were wrong. He threw them out, and took over himself. By June he had dissolved the original junta, replaced every mayor in the country with his own appointees, and announced that new elections would be postponed for at least two years. By July he decreed a state of siege, under which the armed forces were empowered to arrest suspects without charge or right of habeas corpus, to take over private homes and vehicles, and to legally break into homes and offices at night. Then he created special closed military tribunals with the power to sentence subversives to death without right of appeal or eligibility for pardon. Next he banned all political parties. Then he banned the media from reporting any military news that wasn't issued from the palace PR office. He then closed rural provinces to reporters and foreigners, and called up all former soldiers under thirty for the draft. By August there were widespread reports of civilian massacres in the countryside; the death squads were back in action in the cities. And Guatemala was back in the news as America's worst human-rights offender. The general wasn't doing Guatemala's international image any good at all, and the other generals didn't like that. So they'd made their move, but it was premature. The general took once more to the air waves, making his usual speech about God's special plan for Guatemala, the army of

Christ against the Communist subversives, and promising a return to democracy, but refusing, also as usual, to set a date for elections. Then he took the precaution of banning all meetings of more than two persons, and reinforced the machine-gun nests on the palace roof. At the moment, the guns were trained on the schoolchildren touring the place. Some country we'd decided to go bird-watching in.

We descended the marble centre staircase of the palace, facing a handsome wall mural. It portrayed the defeat in 1524 of the Mayan *cacique*, or chieftain, Tecun Uman, by the Spanish conquistador Pedro de Alvarado, founder of the Guatemalan nation. Against a battle tableau illuminated by golden flashes, a glittering Quetzal hovered over the Indian warrior's head. The Spaniard, mounted on his rearing charger, thrust his lance through the native hero's breast. For a scene depicting the birth of a nation, it was a nasty reminder of conquest and subjugation, full of pain and humiliation for the Indians. According to this legend (for the Mayans never practised the distinction between history and myth, and the Spaniards had no scribes there to witness the events), the Quetzal was spiritual protector (*Nahual* in Maya Quiché) of the Indian chiefs. The bird would accompany them on all their undertakings, aiding them in battle, dying when they died. Thus when the Spanish forces arrived in what is today the city of Quezaltenango (the place of the Quetzal), the bird appeared over the battlefield, crying out and pecking at the Spanish interloper Alvarado with its beak. But the Quetzal's magic proved no match for European military technology: mounted on horseback and shooting firearms, the small party of Spaniards decimated 30,000 Mayan fighters in that single day. At the exact moment when Alvarado pierced Tecun Uman, the sacred Quetzal suddenly fell silent and plummeted to earth, covering the body of the regal Indian with its long and soft green plumes. After keeping a deathwatch through the night, the bird that rose from the *cacique*'s lifeless body at dawn was transformed. It was no longer the pure green of jade. Its breast had soaked up the blood of the fallen warrior, and so, too, became crimson, the shade of Mayan blood, as it has remained to this day.

Neither legend nor mural made sense to me. The Quetzal ought to have died along with Tecun Uman. That the bird rose again transformed in the legend shows it as the Mayan version of the Resurrection, influenced by Christianity. However, the Mayan resurrection went further than the Christian. 'Don't you see his breast red as blood, and his arms, green as the blood of plants?' wrote Miguel Ángel Asturias, the Guatemalan Nobel Laureate, in his book *The Legends of Guatemala*. 'Blood of the tree and blood of the animal! He is bird and he is tree! Don't you see the long plumes on his tail! Bird of green blood! Tree of red blood! It is He!'

According to Asturias, the resurrection of the Quetzal embraced not just men, but all of nature, plants and animals. The spirit of this bird leads the entire animate world to salvation. Why, then, I wondered, did the Quetzal portrayed in the palace mural flying over the dying Mayan, wings splayed in agony, already possess its bright red breast – as it appears in nature, but as it appears in the legend only *after* the apotheosis? Maybe this was the confusion inherent in blending the Christian and Mayan faiths.

2

Next day we went to the National Museum of Natural History to see if its director, Jorge Ibarra, could tell us something about travelling in the countryside. Ibarra is Guatemala's best-known living naturalist and a fixture of conservation causes in a country where conservation is an exotic – perhaps subversive – subject. The museum was located in a funky, unmarked, clapboard shed, standing in a yard full of tall weeds, surrounded by a fence of barbed-wire strands on an unnamed dirt road on the outskirts of town. Across the way stood the Air Force Polytechnical School at the edge of a flat and dusty plain, where the cadets marched up and down, back and forth, again and again, in the hot sun.

Through a sliding barn door that didn't slide, we squeezed into a dim and dismal menagerie. The stuffed specimens were crowded into glass cases, which were so crowded together that you had to shimmy sideways through the narrow passages between them to see the exhibits at all. The place was enough to give a naturalist nightmares. It was not that everything was dead, which you expect in a museum, but that everything was so dead, dry, and brittle that you couldn't imagine any of the species displayed there had ever been alive. It was not simply that time stood still there, but that the most important century in the history of natural science had been mummified and entombed: the century of Charles Darwin, Alfred Russel Wallace, Henry Bates, and W. H. Hudson. They had all been drawn ineluctably to the American tropics, where flora and fauna burst forth in dizzying profusion, where nature's cup overflows with life forms. This tremendous dynamism of tropical nature had stimulated their intellects to reach beyond the collection and classification of species, which were the gravamen of nineteenth-century science, into the profound mysteries of natural selection and evolution. Indeed, without the New World tropics, those theories might never have been proposed; the abundance and diversity of tropical life were the key for most of what we have subsequently come to understand about the evolution of life on earth. Yet here one was plunged backwards, through the desiccated bird skins, along the endless indices of pinned butterflies, through labyrinths of Latin taxonomic terminology. Even the tarantulas, the spiders, and the snakes looked as though they were modelling for vast, unread catalogues of species. It took several minutes of silent contemplation before I reached the tentative conclusion that they were actually alive.

Don Jorge was out, but had left a note saying he'd return shortly. Mickey split, suffering an acute attack of photographer's anxiety, caused by entering the close, darkened, inactive room. There were no other personnel or visitors about, so I spent a few minutes studying the most immediately attractive exhibit, a wall poster with Diego Molino's marvellous colour photos of the Quetzal, published by the Banco Industrial. Its text, in translation, read as follows:

The Quetzal, the most beautiful bird that exists in the American continents, belongs to the Trogon family. The iridescent colour of its plumage appears green or blue according to the changes of daytime light. In our country, it lives in the mountainous, subtropical, humid regions of the departments of Quiché, the Verapazes, Huehuetenango, San Marcos, and Suchitepéquez. The vegetation of the territory it inhabits is quite dense and rich in humus. In this habitat, the Quetzal searches for an old tree trunk situated in a tiny forest clearing to make its nest. There, between February and April, the hen lays one or two eggs. Both the hen and cock take turns during the eighteen-day period of incubation. The male Quetzal enters the nest, always leaving his beautiful tail plumes outside so as not to injure them. The female doesn't have this problem, for her tail feathers are very short. After the birth of the nestlings, their parents feed them with worms, insects, and larvae. The adults will eat forest fruits. The young can fly twenty days after birth, and abandon the nest to cut freely through the skies of our nation. If the Quetzal is confined in a cage, it dies. The Quetzal cannot live in captivity. For this reason, it is the emblem of our liberty.

Ibarra arrived a few minutes later with his shy young female assistant. He was a short and dapper man, a David Niven double, with the neat clipped moustache, and moist blue eyes. He wore a baby-blue cardigan sweater with a crocodile on the breast. They were carrying stacks of *Historia natural y pro natura*, the occasional magazine that Ibarra writes, edits, designs, sells ads for, and from the looks of it, personally delivers, to raise conservation consciousness, not to mention funds for his museum. He said he'd been applying for government funding for a new building to house his collection for twenty years. But every time the National Congress appropriated the money, the construction materials were mysteriously delivered to sites where certain military officers were building themselves new homes. It's not the kind of thing you blow the whistle on in Guatemala, the director indicated.

'But things look a little brighter under the present regime,' Ibarra said. 'At least the Minister of Agriculture is familiar with the concept of conservation. We've even broken ground for our new building.'

I wondered how many times in the past things had looked 'a little brighter', and an image flashed across my mind's eye of poor old Don Jorge Ibarra, slamming shut the museum door, dousing the lights, and diving behind the glass cases, as the gunshots of another coup started at the Air Force Polytechnical School across the road.

While his assistant went to prepare coffee, Don Jorge led the way to his private office, and assumed the director's chair behind an immense glass-top desk without a scrap of paper on it: he looked as though a great tidal wave were about to knock him over. The office had a bookcase with a sign that read, '90 books which mention the name of the director of this museum'. Over the bookcase hung poster-sized blow-ups of Ibarra's father and grandfather in dour black frock coats and stiff Edwardian shirt collars. Another sign described them as 'Pharmacists and men of science'. The walls were plastered with degrees, awards, proclamations, and certificates – one, I noticed, was a calligraphy letter in Italian, congratulating Ibarra for his conservation efforts. It was signed by the king of Italy – though, surely, Italy hasn't been a monarchy for a long time. Another scroll bore the title, 'Order of the Quetzal', Guatemala's highest civic honour.

Ibarra accepted a cigarette, and got right down to bemoaning the status of Guatemala's national bird. The original range of the Quetzal, he said, covered 30,000 square kilometres of the country's highland cloud forests. By 1981, it had been reduced to 2,500 square kilometres. 'In most places, the Quetzal has been wiped out only in the past fifteen years,' he said. 'And why – because of shooting or poaching? Maybe a little. The Indian lives like an animal. He has no culture. He shoots whatever moves. What doesn't move, he chops down with his machete. What he can't cut, he burns. That is the real problem here – cutting and burning of forests in elevated habitat for firewood and *milpa*, the Indian's cornfield. Experiments have shown that as the forests are cut down, the mean temperature of Guatemala has risen. This change of temperature adversely influences the cool, upper forests needed to produce the Quetzal's food. The Quetzal's favourite foods come from the *Lauraciae* family of small fruits, of which the avocado is the outstanding member. The Quetzals eat what we call the *aguacatillo*, the miniature avocado, as well as other types of small fruits. But these trees require high elevation, cool temperatures, and a great deal of moisture.

'The government isn't interested in reafforestation,' he went on. 'If you go to the government and ask, they will tell you that Guatemala adopted a model reafforestation law in 1935. They will say that ten million trees have been planted since the reafforestation law went into effect. Sure, they planted pine trees. But they never did one thing to care for them. They didn't teach the Indians how to cultivate the seedlings. It was propaganda, that's all. I can state definitely that Guatemala has fewer trees today than when reafforestation became the law almost fifty years ago. The solution can only be planting trees as the forests are cut down, along with the education of the Indians. But I am not very hopeful. We are running out of time. We predict that the Quetzal will be extinct by the year 2000. What can I do about it? I am only one person here, alone. Guatemalans don't understand conservation. They don't care. Their minds have not been developed to appreciate natural beauty.'

He pointed behind his desk to an architect's drawing: a bleak stone obelisk, with a lily-white couple standing in the foreground holding hands. 'You see this? When the Quetzal is eliminated, they'll build this stone monument as a memorial. Too bad – it's even an ugly sculpture.'

An authentic voice of Central American pessimism.

Ibarra was dubious of our chances of even observing the Quetzal; it wasn't only the rainy season. 'You have to realize, in this country there's always danger involved in bird-watching,' he said. 'You put your binoculars up towards the trees, and people don't know what you're doing. They become frightened. If you go on to someone's property to look at a bird, they don't ask any questions, they shoot you first.'

Was it true, then, as we had heard, that a well-known Guatemalan naturalist had actually been assassinated?

'I don't know about that,' Ibarra said after mulling it over. 'But there are guerrillas. The government says the countryside is pacified right now, but sometimes the government tells us that an area is calm when it really isn't calm. There's no way of knowing. The way you propose to go, without documents or safe conducts, I would say this, in general. The Verapazes? You should be all right. Quezaltenango? All right, but the Quetzals are gone from Quezaltenango. Quiché? We don't know the situation in Quiché. I would stay out. Besides, the forests have been cut. Huehuetenango? Chichicastenango? Don't go there. Too dangerous. *Mucho conflicto*.'

He was disparaging, too, of the Biotopo del Quetzal. 'Let's face it, the Biotopo is a tourist vehicle,' he shrugged. 'No scientific or conservation work is accomplished there. The very idea of establishing a wildlife refuge for people to visit is counterproductive. People drive there in cars, step out, and say, "OK, where's the Quetzal?" Of course, as soon as people started invading their habitat *en masse*, the birds left. Believe me, you won't see Quetzals there.'

Ibarra had been urging the government, as an alternative, to establish a more remote refuge at a place called Chelem-ha, an unblemished mountain he described as north of the Río Polochic, high in the department of Alta Verapaz. The mountain was owned by his friend Alfredo Schlehauf, a coffee grower of German descent. 'Chelem-ha is an extraordinary place, completely wild,' Ibarra waxed enthusiastic. 'You can't reach it by car, there's no road, only a dirt path. I've been there four times with Don Alfredo Schlehauf, who is an outstanding gentleman and true friend of conservation. The last time I went we had the miraculous experience of sighting six Quetzals all at once. It was in April, the beginning of the nesting season. We had gone up to Chelem-ha from Don Alfredo's plantation, Finca Remedios, eighteen kilometres away, partly on horseback, partly on foot. Suddenly, six Quetzals flew overhead. Three landed in one tree, three in another. All males. They had passed over us in such a way that only their backs and tail plumes were visible. The greens were very iridescent in the sunlight. Then they turned when they perched, so that we saw the bright crimson bellies. It was out of this world. In that moment, I was thinking, This is the greatest reward of my life as a naturalist.'

It sounded good. I had explained the trouble my partner and I were having procuring safe conducts from the palace, and now I wondered out loud if Chelem-ha might prove the best site for our Quetzal watching.

'Why not? Good idea,' Ibarra agreed quickly – perhaps a bit too quickly, I thought, for Chelem-ha was clearly his pet project, and you had to gauge anyone who had survived so long in Guatemala as a shrewd operator. Nevertheless, Don Jorge volunteered to introduce us to his friend Schlehauf, and promised to obtain permission for us to travel to Chelem-ha. Just like fat Escobar at the palace, Ibarra called in his assistant to telephone the German. The incapacity of male Guatemalans to dial the telephone for themselves seems to be endemic. While she did the telephone duty, Ibarra muttered nervously through the labour pains of the project: 'It would be better if you met Schlehauf yourself and spoke to him personally . . . yes, yes, that is what I'll say . . . I wonder if the road to the *finca* is negotiable this time of year. Sometimes it turns to mud. You may need a four-wheel-drive vehicle. I have a friend with a rental agency. Maybe I should call him, too. Or you could go with Don Alfredo – that is, if he's travelling to the *finca*. You had better talk with him personally,' etc., etc.

Naturally, Schlehauf wasn't at home.

Ibarra promised to reach him and get back to us at our hotel, but he never did, and neither did Escobar. Maybe their secretaries had gone home for the weekend.

3

The Guatemalan national medallion, which also features the Quetzal, stood atop the pink wedding-cake building that houses the central post office. Within a frilly stucco wreath were crossed muskets, as well as crossed cutlasses, and within the weapons a scroll proclaiming, 'Liberty, 15 September 1821', the date Spain conceded Central America independence without a fight. The bird perched on the upper-right-hand corner of the scroll. Its tail plumes flowed gracefully around the curved outer edge of the scroll, and ended in green curlicues among the cutlass handles. What a difference from the rapacious birds of prey of so many national emblems! – imperial warriors of the skies and all that, I admit, but also greedy nest robbers, home wreckers, and blood gluttons. This Quetzal didn't look like it would harm the hair on a fly's head. It was a childlike representation of freedom – a freedom hardly troubled by the muskets and cutlasses of political struggle in the background, the natural, idyllic, ethereal liberty of that handsome, free-flying creature.

Under the sign of the Quetzal, the black market was holding its trading session. It was a fitting site, inasmuch as the Guatemalan currency is also called the quetzal. Men and women of every age, size, and colour were rushing up and down the pavement, feverishly offering prices to passers-by, much as one imagines the American Stock Exchange long ago in the days before it got its own building and was known as 'the curb'. The Guatemalan government claims pride in the fact that its quetzal is the only hard currency in Central America. It has been maintained at the official exchange rate of 1 quetzal = 1 US dollar through all the financial crises and political turmoil of recent decades, but you'd never guess it walking past the post office.

'Pssst! – over here,' called a fat woman running a postcard stand. 'Quetzales. Quetzales.'

Before we could even reach her, a smoothie in a black knit shirt and dark shades took me by the elbow and manoeuvred me downstream through the swarming traders. 'You have dollars? I'm paying 1.26 quetzales. The rate's going down tomorrow. You'd better change now.'

Funny, how you always seem to hit the black market just at a moment when the dollar's sinking.

Then a red-turbanned Indian – an Indian Indian – with a neatly trimmed beard and clipped English accent intervened. 'Please,' he said politely, 'I like America very much. You would perhaps help a fellow from Bombay earn airfare by changing with me, yezz? I will pay you 1.27. You see, I would like to go to America very much.'

'One fifty,' demanded Mickey, getting into the swing of things. He added, in his abysmal Spanish, '*Nada más!*'

'One thirty over here,' whispered someone from behind.

'I'll top that. Over here. Change with me,' said someone else.

We were in danger of being torn to shreds as offers and counter-offers whistled past our ears like tracer bullets when suddenly a dust storm blew up in our faces. Out of its centre, as if out of a spaceship, stepped the raggedest-looking kid ever to hold a seat on any exchange. There was dirt smudged all over his round, brown face, dirt caked under his fingernails. He wore a torn and filthy T-shirt that read, 'Have a Pepsi today'. There were kneeholes in his shiny pants. He was barefoot, and his jet black hair had been shaved to crew length, probably to protect against lice.

'One fifty,' announced this Boy Wonder of the Black Market, and with total self-assurance cordoned us off from the other traders by going from one to another, earnestly pushing them back. Then he took my hand and marched me down the side street, vehemently urging us on with cries of '*Vamos! Vamos!*'

'Hold on a moment, *chico*,' I said. 'Where are you taking us?'

'To the office of my boss.'

'Who is that?'

'The Chinaman!'

'He'll pay us one fifty?'

'One forty-two.'

'But you said one fifty ten feet back.'

'That was then and this is now,' he answered, not batting an eyelash. 'The market moves fast. What are you changing?'

'Dollars.'

'Ah, dollars.' His tongue licked his lower lip. 'Cheques or cash?'

'Cash.'

'That is well. Cash is very good. You get more for cash. I get more for cash – my commission, you know. One forty-five for cash, OK? *Vamos! Vamos!*'

'Let's get out of here,' said Mickey.

'Look, you won't get a better deal on this street, I swear,' implored the Boy Wonder. 'You heard what they were saying, the dollar's headed down. Guatemala just got a twenty-million-dollar American loan. The market's flooded. Tomorrow, no one buys dollars.'

'One forty-seven or goodbye,' I responded. It felt ridiculous, not to mention slimy, to be discussing exchange rates with a scruffy kid: he couldn't have been older than twelve, and the impropriety of his holding my hand while wheeling and dealing never occurred to him. But the reality of black-market transactions is that the American with a few hundred bucks in his pocket feels so vastly, overwhelmingly, scandalously richer than nine tenths of the population that his will to hold out for a higher price is compromised from the start. Squeezing another two or three quetzals out of the deal made you feel like a child pornographer.

'OK, OK,' the Boy Wonder relented. 'For you, a special price. I'll see what I can do with the Chinaman. *Vamos?*'

'*Vamos.*'

The shop was only a few steps farther down the side street. It was like a thousand other shops of the Oriental diaspora – clean, neat, well organized, and understocked. The tile floor was flawlessly mopped, and smelled strongly of disinfectant. The imported trinkets, mostly cheap electronic gadgets, were arrayed under a long, glass-top counter, while the gym bags, sneakers, soccer balls, and the like were bagged in plastic on the wall shelves. A pretty Asian woman leaned over the counter, gazing into her private eternity of boredom while the flies buzzed around her head. Through an open curtain at the rear, you could see her children, so intent on their stock-clerking chores that you had the impression of tiny grown-ups. Stationed on a metal frame settee, the grandfather stroked his long white goatee and blew cigarette smoke in precise, thin streams, out into an air of dreamlike stagnation. No one moved when we entered, not even to look at us: it was like entering a wax museum.

At the rear, the Chinaman did business behind a small school desk. He had a broad smile and calm manner. He asked where we were from and admired Mickey's cameras. He offered us cigarettes from an open pack of Marlboros, lit mine with a Bic, and gave one to the Boy Wonder, who sheepishly retired to the far side of the grandfather before mumbling a few words about having offered us 1.40. The Chinaman kept his cool. He said it was simply out of the question. He didn't set the rate. That was done by a pool of big shots from the commercial trading corporations, the government import authorities, the banks. He was just a little guy, a cog in the great wheel of currency corruption. He needed dollars to pay for the goods he imported, that was all. He could pay us 1.37, tops. He took a long drag of his smoke, then added coolly that we could cut the Boy Wonder's commission out of the deal. He'd lied to us, hadn't he? In that case, he would pay 1.38.

The Boy Wonder stood looking at us with a melting expression. Every particle of dust and dirt on his frail frame implored us not to take the bread out of his mouth.

It was our turn to relent, which we did.

The Chinaman opened the bottom drawer of his desk, and removed neatly banded bundles of new quetzal notes – lavender fives, brown tens, blue fifties, red hundreds. They were the most attractive currency I had ever seen. Each denomination bore the engraved portrait of a different Guatemalan tyrant, and all had a svelte green-and-scarlet image of the Quetzal sailing across the top, its tail plumes crowning the tyrant's head. The bills were bordered by intricately engraved renderings of Mayan glyphs, so colourful and complicated they looked like miniature tapestries.

The Boy Wonder tried to swipe a five-quetzal note off the top, but the Chinaman quickly snatched it back, then reached into his pocket and handed him a crumpled note. It was a fifty-centavo note, half a quetzal. I asked to see a new one of the same denomination. The engraving was not of a tyrant, but of Tecun Uman, the Mayan chieftain. I gave it to the Boy Wonder, who grunted his appreciation, then slipped out the shop front while we were concluding the deal. We never even got to ask his name.

4

Friday afternoon: the engines roar, the horns blare, the city empties early. By five o'clock, Diagonal 12, which is the name of the main commuting thoroughfare, was clogged with private cars, motor bikes, trucks, and smoggy buses. There were lots of diesel engines, but very few functioning mufflers. The pollution sat in a colourful pink-and-grey haze like a cotton candy cloud, because Guatemala City is situated in a valley below rising volcanic peaks, and the air doesn't want to move at all. Traffic was backed up for blocks at the intersections. The drivers sat in air-conditioned comfort with their black glass windows rolled up. Workers hung off the sides of the overcrowded buses. Truckers looked as if they bathed in dust.

Meanwhile on the sidewalks, a euphemism for the dirt paths worn along the street sides, Indians streamed along in indescribable confusion, like a horde of leaf-cutting ants marching along the jungle floor in endless files, carrying gigantic titbits of vegetation on their backs. The Indians appeared to be overcrowded and under-employed and, mostly, tired. You don't know what it's like to be poor and tired until you see those round, stony, bereft Indian faces, moving zombie-like through the dusty streets and markets. Flocks of barefoot Indian kids swarming around cars to beg a penny. Arthritic old Indian women carrying their unsold loads of bananas and baskets of beans, chillies, or cornmeal, on their heads. Listless mothers lugging infants on their hips, staring sullenly at the black glass windows of a car in the vain hope that the driver will take pity and give them a coin to buy food for their babies. Indian men trudging along under loads of a hundred pounds and more, like weary beasts of burden. Crippled, blinded, stunted, disfigured, ragged urbanized Indians weaving through the traffic lanes trying to sell Chiclets or afternoon papers, or charcoal grills made out of old auto wheel rims. You didn't see any Indians behind the wheels of those BMWs and Chevy Blazers. The Indians trudged, or took the bus if they had the fare.

The population of Guatemala City, I suppose, more or less reflects that of the country as a whole. For the capital city grew by in-migration from the countryside to its present sprawling size of about 1.5 million. But demography isn't an exact science in Guatemala, maybe not a science at all. Demographers aren't sure if the Indians don't cooperate with the census takers, or if the government manipulates the statistics, or both. In any case, the indigenous people, as they prefer to be called, are severely under-counted. There are said to be twenty-two Mayan dialects spoken in Guatemala, each dialect representing a major tribe; but what portion of the total Guatemalan population of about 7 million is Indian, no one says with any certainty. Some analysts state merely that the indigenous peoples constitute a majority, and let it go at that. Others claim that as much as 70 per cent of the Guatemalan population is basically Indian. The government says 38 per cent. But just who is an Indian; whom should the population counters count – only full-blooded Indians? Only those who still wear native costumes? Only those who answer, 'Yes, I am an Indian'?

The government says 38 per cent, but the government estimate is certainly too low. Needless to say, the government and the military aren't run by Indians. They are run by Ladinos, which is the term for those of purely Spanish, or mostly Spanish, or partly Spanish, or even imagined Spanish, descent. Ladinos can be dark-skinned, or pale as a Swede. They may be broad-nosed and lack facial hair, or blond with blue eyes, or anything in between. Racial characteristics aren't useful in defining the Guatemalan population, but economic, social, and even personal characteristics are to some extent. Ladinos are more likely to be middle class, live in houses, own cars and motor bikes. They speak Spanish as their first language. They own most of the real estate, make most of the money, and commit most of the crimes, economic and violent. The Indians are more likely to walk, and carry things on their heads and backs as their Mayan ancestors did. They are more likely to go home to their rural villages each year when it comes time to plant maize in the family *milpa*, or farm. Many have only a rudimentary knowledge of Spanish. The majority of Indians are illiterate, and most never attend school past the third grade. They are less violent and less prone to anger than Ladinos. They are generally less aggressive, more religious, and, of course, poorer. In the end, who is an Indian and who a Ladino is largely a matter of cultural attitude. It's who you believe you are that counts. The Ladinos are the culturally dominant group in Guatemala, much as whites dominate in

South Africa. Whether Ladinos constitute 20 per cent or 60 per cent of the population really doesn't matter. What matters is that Ladinos run the census, and every other public institution.

By twilight, around 7 p.m., only stragglers remained out. For a moment, Guatemala City became excruciatingly beautiful: the molten sun setting in that roseate fluff behind blue mountains; the refreshing trace of cool pine scent on the damp evening breeze; and a host of Boat-tailed Grackles kibbitzing overhead as they made for their night roosts. Then darkness fell with a thud, and the whole city went eerily silent. In the foothills ringing downtown, you could see firelights flickering in the shantytowns and smell the charcoal burning. But neither Indian nor Ladino was abroad downtown. There were no window shoppers, no young lovers, no strollers out taking the air under the tall banyans on Avenida Reforma. No one was out for a Friday-night car cruise, and there weren't any of those small clans of men on the corners, talking, drinking, and making music, as there are in other Latin American cities. There weren't any diners going into the restaurants, most of which were already closed, no kids out at the movies, no hookers on the streets. *Nada.* It was as if they'd rolled up the carpets and opened all the panther cages at the zoo.

We walked around looking for a place to eat, and after finding a Kentucky Fried Chicken franchise still open, walked around some more, trying to find out what happened to all the people. The ruins of the buildings destroyed by earthquakes stood in the shadows like heaps of dinosaur bones. The last big one was in 1976, when 30,000 lives were lost and hundreds of buildings wrecked. But it wasn't fear of earthquakes that kept the residents shut away at night. We began to feel edgy about being out ourselves, started looking over our shoulders, peering into the dark, to see if anyone was hiding in the shadows. The mood of predation and fear was palpable. As Mickey and I sat over coffee on the patio of the only open restaurant we could find, an unmarked van pulled up across the street. Four men emerged to start unloading long, narrow objects in the shadows of a ceiba tree. My night vision is doubtful in any case, but I did notice the men gesturing towards the patio and looking around, and I realized we were the only ones sitting there. When I saw them scurry across the street with their dark packages under their arms, my heart knocked up against my tonsils.

'Here come the death squads,' my partner said light-heartedly.

They turned out to be a *mariachi* band. The ominous things under their arms were guitars and trumpets. Perhaps the *señores* would like a tune?

5

On Sunday, we followed our noses to the Guatemala City dump to watch the Zopilotes, or Black Vultures. That inglorious establishment wasn't hard to smell. The sour stench was pervasive for blocks all around, until we spotted the birds soaring lazily in a threatening sky full of rolling zinc- and obsidian-coloured thunderheads.

A whole *barrio* had grown around the dump, a neighbourhood of human scavengers. Their shanties began at the very edge of the dump and circled out, so close together you'd have thought it a prestige address. In the stony, rutted, unpaved streets, more like dry riverbeds, the small kids played catch, the teenagers soccer. Mongrel dogs yapped and fought over turf or scraps. Unappetizing hens grovelled in their dust baths. A few women walked to church dressed in frilly pastel shifts, Bibles cradled lovingly in their arms: a note of serenity in the filth and tumult of the slum. But for most who lived by the dump, Sunday was a workday, as gruesome and humiliating as all the others.

A constant stream of people flowed up from the vast pit below, advancing along steep, well-worn paths. They walked slowly and bent over, loads of salvaged materials piled high on their backs and attached to their bodies by waist and forehead straps. The older ones, the lame, and the blind leaned on their children as guides or on staves, if they had no children. Their bare, gnarled feet felt along the ground for toeholds. Their eyes never left the ground, their faces locked in concentration. They passed the car we'd rented without so much as a glance, as if the slightest movement might upset their precarious balance and send them toppling over. At least mules have the perverse pleasure of refusing to move now and then, even if it costs them a beating. But these were men and women, and they had no choice but to bear their burdens: no one would beat them if they fell; no one would help them up either. A second stream of 'empties' filed back downhill, their bodies doubled over on themselves like hairpins from the routine posture of their work.

We circled the zone for a way down into the dump itself, where the birds were. A steep street provided access for the garbage trucks. Not dump trucks, which make things easy, but large stake-bodies, so that the refuse had to be shovelled out by crews standing hip deep in dross. Along an adobe wall running next to the road, the crewmen stood in filthy tatters of grey uniforms, lost in a cloud of flies, taking a break. One of them approached the car as if to make official inquiries. But as he drew closer, motioning with his hand for us to stop, I noticed the bleared eyes and the stagger. I glanced back at the others, saw the bottles of rotgut, and realized they were drunk. Sprawled on the broken glass and twisted tins, asleep or passed out with the flies in their eyes and their mouths, while others leaned against the adobe, heads sunken against their meagre chests. A few stared at us with hopeless indifference or plain disbelief. The sight of visitors, especially Yankee visitors, must have been a rare one. As the man weaved towards us, I slowly swung the car around him, a timid bull round a tipsy matador, and kept going. He, however, leaped forward, lashing out with his foot, and smashed the tail light of the rented car.

We tried another access to get closer to the vultures, concentrated in a thick black blanket a hundred yards in front of us. It put us slightly above the dump, at the edge of a sheer drop, but one of only about thirty feet. Indian women went back and forth there, some carrying refuse to the edge and heaving it over, the rest climbing out with the stuff they'd collected from the rubbish mounds farther down, near the trucks. This was apparently the dump's high-rent district: the row of sheds doubling as homes perched above the pit was easy to reach, and the more specialized scavengers stored their stinking treasures here. The carriers dropped their loads, women and children sorted and stacked. Each shed was filled with a particular type of discarded matter. One dealt exclusively in glass bottles, another in tin cans, a third in wooden scraps, a fourth in corrugated metal, a fifth in cloth rags, a sixth in rubber tyres, a seventh in burlap bags, an eighth in plastic, and so forth. It was a Central American garbage mall, with dozens and dozens of garbage boutiques. The only thing missing were shoppers with a taste for trash.

Directly below, an indelible scene spread out like a ghastly mural of hell's mess hall: hundreds of men, women, children, vultures, dogs, and rats, all gathered round the open rear ends of the yellow garbage trucks, waiting for the raw slop the crews were shovelling off. The competition for anything edible was intense. Zopilotes dive-bombed, carrying their prizes off a safe distance before digging in. The humans didn't bother with such niceties. They ate directly from the mounting piles under the truck gates, or climbed right in to the truck beds and rummaged through the debris. Fights broke out, and people wrestled in mounds of wet, smouldering garbage. The dogs got the short end of it. The brazen vultures had little trouble driving the dogs away from their finds, and the men, too, picked up any handy stick or rock and lit into the cringing canines. One man went after a dog with his belt, and administered such a savage whipping that the howling mongrel stumbled, then keeled over. Then the man grabbed a stone and bashed the dog's brains in. The Zopilotes didn't waste an instant before claiming the warm body for themselves.

Appalled and riveted, we shut the motor and got out of the car. As we did so, a boy of maybe ten ran up to Mickey, who was wearing his cameras, and said, 'Be careful. They tried to kill the last photographer who came here, to steal his camera. And his wasn't as nice as yours!'

We thanked the *muchacho*, and introduced ourselves. His name was Leonel, but we called him 'Ratón Mickey' because he wore a clean white Mickey Mouse T-shirt. Spotless red gym shorts that grabbed at his chubby thighs, white tube socks, low-cut black sneakers, and a baseball mitt on his left hand completed the attire of this goodwill ambassador. In addition to being exceptionally well washed, he was handsome, plump, curly-haired, spunky, bright-eyed, and loaded with charm – something of a miracle of parenting, you would think, given the neighbourhood. He volunteered to guide us; the only charge, it was immediately clear, would be answering his questions, a small enough price to pay. He was curious about everything. Where did we come from? What were we photographing? How did one say 'Adidas' in English? (He looked slightly crestfallen when I explained that 'Adidas' was a name, the same in English as in Spanish.) Which did we think better, baseball or soccer? The Los Angeles Dodgers or the San Diego Padres? How did we call the Zopilotes in English?

'Vulture,' I told him. 'Vul-choor.'

The strange syllables rolled around inside his fat cheeks and tumbled out into the noxious air. He giggled, 'What does it mean?'

'Nothing, it's just a name, like "Adidas". What does "Zopilote" mean?'

'Nothing,' Leonel said. 'It's a name also. But around here we call them "bachelor eagles".'

'Why is that?'

'I don't know, it's just what we say. I think it's slang, because you can't tell the difference between the man and the woman Zopilote.'

'Do you like Zopilotes?'

'Oh, yes, I like birds – and these are the only birds I ever get to see,' he responded ingenuously. 'Last year I wanted to see their nests, so I walked down there, far back, to the caves where nobody goes.'

'What month was that?'

'It was, let me see – April.'

'And did you find their nests?'

'They don't really make nests [he was absolutely correct on this point], but I saw their eggs. They lay their eggs in the caves back there, on the ground.'

'How many eggs did they have?'

'One egg for each pair. The eggs were all white, and very large.'

'Excellent report. Did you see their babies, too?'

'Some, yes, of course.'

'How did you know they were babies?'

He thought for a moment, then said, 'Because they were small, more grey than black, and the parents were feeding them, too. That is how I knew.'

'You made quite a study of the Zopilotes, didn't you, Ratón Mickey?'

He bubbled with pleasure under the compliment. 'In the Estados Unidos, you have Zopilotes, too?'

'Yes, but not as many. Most of ours are a different colour. Ours have red heads.'

'We don't have the red-headed ones here in Guatemala. All ours are black. I'd like to see the Zopilotes with the red heads. Is the Estados Unidos far from here?'

'Not very. About an hour and a half by jet.'

'By jet!' He swooned at the word. 'I'd like to go in a jet so much! Listen, you wait right here, I have to go get something, OK?'

'OK.'

Mickey set up his tripod at the edge of the promontory and started clicking. By this time, a horde of kids had completely encircled us, mainly to ogle the camera and its foreign owners. They stuck their tiny faces and waved their little hands in front of the lens, and struck silly poses with their sisters and their friends. Someone else threw a rock. In the street near by, two men were rolling in a drunken brawl. A chicken on a chain raced into the clump of children, with a dog in hot pursuit and a man chasing the dog to save the chicken. In an instant, the animals were completely tangled up in the legs of Mickey's tripod, with the man beating the dog, the dog yowling, the chicken feathers flying. It was all we could do to save the equipment. The kids scattered, but in the commotion a redheaded little girl slipped over the edge, and we had to jump down and rescue her from the garbage. By the time we brought her up, the kids were smearing their fingers all over the camera lens.

But our little bedlam was mild compared with what was going on down in the pit. One of the ragpickers had left off his garbage combing, moved towards us through the trash heaps, dropped his pants, and unceremoniously commenced to defecate. No sooner had he dropped his load than the Zopes rushed to that spot as though the most marvellous concoction had been set before them. Zopilotes are great shit-eaters. Half a dozen of them gobbled the stuff down in a matter of seconds. It was evident now from every point of view that the dump was not a birding hot spot.

On our way back to the car, Leonel popped up again. He had run home to fetch his little school atlas, which he held out proudly in his baseball mitt. He wanted us to show him on the map where we were from. I pointed to New York, Mickey to Chicago.

The boy gazed quietly at the map for a moment, then said, 'Sometime, I'd like to go to the Estados Unidos. I would like to study to be' – he paused to think of the most impressive-sounding title – 'a car mechanic. Do you think I could do that in the Estados Unidos?'

'Why not?'

'Where should I go – to New York, where you're from?'

I didn't know what to tell him. The world is wide and life is funny, but not that wide, not that funny. A kid from the Guatemala City dump would need extraordinary willpower just to get himself to the other side of town. He was a plucky kid, though, and he questioned everything and wanted to find out for himself. It occurred to me that he was probably just the kid who would some day ask the wrong question and end up in the dark maw of the death squads. I didn't want to thwart him, nor encourage him falsely, nor, most of all, lie to him. 'Go to the United States,' I told him. 'As soon as you get old enough, go. Walk through Mexico, if you have to, but get out of here – you understand, Ratón Mickey?'

'Yes.'

'It won't be easy for you there. Jobs are hard to find. School costs money. There's prejudice, and many poor people, in the United States too. But you go. Don't go to New York, or Chicago. Go to Miami or Los Angeles. Many people there speak Spanish; they'll help you. O K?'

I drew circles around the two cities on his map, so he might remember. He seemed satisfied. He asked, 'Did you get good photos of the Zopilotes here?'

'Plenty.'

'There are more Zopilotes here now than there used to be.'

'Is that so? Why, do you think?'

'I don't know,' he said. 'But I think the Zopilote has a great future in our country. It eats the dead things. And here we have more and more dead things all the time.'

6

In a surplus pannier case I carry over my shoulder are maps, maps, and more maps. Maps of Central America. Maps of Guatemala. An archaeological map, in case we journey near some famous ruin. A road map, in case we ever get out of Guatemala City. A topographical map, to tell us whether we're up or down. It's easy to get disorientated in Central America. It's easier to get lost. What isn't easy at all is finding the Resplendent Quetzal, as ornithology officially names the bird we have travelled thousands of miles to observe. For months back home, like some weird religionist, I'd spread my maps on the floor and walls of my library and spend evenings getting myself located in quiet reverie. The Quetzal's range actually extends from southern Mexico through western Panama in mountains of 4,000–10,000 feet in elevation. As with other highland species in Central America, the Quetzals were separated at some point in geologic time into northern and southern populations by the stretch of lowlands that covers parts of southern Guatemala, Nicaragua, and northern Costa Rica. The two Quetzal races evolved locally in these sectors, and are now considered separate subspecies. Our quest was for the northern, or Guatemalan, Quetzal – the Quetzal of myth and legend.

Looking at my maps I could see that we'd be venturing straight for the heart of Central America, the geologic and biological heartland, though not the literal centre. The isthmus of Central America actually begins in the southern Mexican state of Chiapas, which was part of Guatemala until 1842. Then it heads south through Guatemala, the largest country in Central America. It twists eastward through Honduras and El Salvador, then humps through Nicaragua. The neck starts to narrow in Costa Rica, and finally piddles into a thin Panamanian umbilical cord connecting to South America.

But Central America is not one solid, stationary land mass; rather it is a kind of loose, unstable coalition of mountain chains. Up until the mid-Miocene, 23 million years ago, the northern and southern continents of the Western Hemisphere weren't connected. There were seas and island chains between them. Then the two continental land masses started moving closer, as the immense tectonic plates of strata that underlie the surface of the earth shoved up against one another. In that geologic jostling, the Central American sierras formed. Where there were faults, holes, canyons, left by the uneven movement of the plates, the volcanoes rose. Where the seas gradually withdrew, the lowlands are today. Until the time the Central American isthmus formed, North and South America were completely separated. Animal and plant forms on the two land masses developed in isolation from each other. But when Central America happened, there was a new land bridge between north and south. For the first time animal and plant forms could wander to the other continent, even settle there if they liked the neighbourhood. And that's just what they did. Northern species migrated south. Southern species migrated north. It was probably the greatest diffusion of life-forms ever. And more new species than anyone has ever been able to name began to evolve in Central America itself, where the weather was good and the landscape lush.

And among these new species were the trogons, the family of birds of which the Quetzal is the most esteemed member. No trogon has ever been found outside the American tropics. No Quetzal has travelled farther north than southern Mexico, nor farther south than Panama. Guatemala, with its towering spine of high mountains, made a perfect headquarters.

The 1975 map from the National Geographical Institute of Guatemala provided inspiration as well as terrestrial guidance. On it I marked the names and dates of some naturalists who went before us in search of the Quetzal. The first was a nineteenth-century English explorer, Osbert Salvin, whose awesome forty-volume *Biologia centrali americana* remains to this day the most complete catalogue of American tropical species. When Salvin set out for Cobán in the Guatemalan highlands in March of 1860, there was still some question as to whether the Quetzal actually existed. In the 1830s, a French traveller named Delattre had reported sighting a Quetzal while visiting the mountainous forests of the Verapazes. Clamour built among European collectors to own a specimen of this exotic new creature – it was the golden age of stuffed birds. But Salvin was really the first to go to the neo-tropics and observe the species first hand, *in vivo*. The scientific expedition of those days constituted what we now might call a hunting trip: Salvin left, 'determined, rain or no rain,' as he vowed in his journal, 'to be off to the mountain forests in search of Quetzals, to see and shoot which has been a daydream for me ever since I set foot in Central America'.

Salvin soon discovered that even the dogged determination of John Bull was no match for the Guatemalan weather: 'Rain all day and every day is what one must expect to encounter on visiting Cobán,' he recorded drearily. 'Morning after morning brings no change for the better.' He had based his expedition in Cobán because of the town's reputation as the centre of a lively skin trade in animal specimens. While incessant rains prevented his departure for the back country, Salvin kept himself occupied in the town with a great Victorian biological shopping spree: 'A mere hint at what branch of natural history one has a leaning towards is sufficient to bring in specimens in an almost unbroken stream,' he wrote, cheered up. 'Boy follows boy, till one hardly knows which way to turn to stow away the spoils in the shape of birds, snakes, lizards, toads, frogs, etc., and no small amount of time is occupied in paying those young rascals (for they all try to cheat) for their captures.'

Salvin failed to describe exactly how a sophisticated European dealt with these 'young rascals' – illiterate natives outside the money economy – or derived fair prices for their specimens, which would be interesting to know. What did a Guatemalan toad go for in those days? Nor did Salvin admire the Indians at all. His journals go on at length about how the natives cheated him, lied, drank *chicha* (undistilled corn liquor), and most of all, refused to travel in the rain, the lazy bastards. When he finally forced them into action ('the period of my stay being limited, idleness cannot long be endured'), they lagged behind, made excuses, and subverted the brisk daily goals of march he had set for them. When all else failed, they got lost. Sir Osbert noted: 'As no one seems to have a very clear idea of the road, I, compass in hand, undertake the direction of affairs.'

Nevertheless, at length Salvin came within striking distance: 'On entering, the path takes the unpleasant form of a succession of felled trees, which are slippery from recent rains, and render progress slow. My companions are ahead, and I am just balancing myself along the last trunk, when Filipe comes running back to say that they have heard a Quetzal . . . I immediately hurry to the spot.' Momentarily, a splendid male Quetzal settled on a branch within view. Salvin was in rapture watching the bird bob for berries 'with a degree of elegance that defies description'. He points out that while a hummingbird requires favourable position and light to project its charm, 'this is not the case with the Quetzal. The rich metallic green of the head, back, and tail coverts reflects its colour in every position, whilst the deep scarlet of the breast and the white tail show vividly at a distance.

'Unequalled for splendour among the birds of the New World,'

Salvin raved. (It's interesting to note how Salvin manages to compliment the New World species without challenging the beauty of Old World species, although any fair observer studying the matter would have to conclude that comparing the birds of Europe to those of the American tropics is like comparing an English matron to a Brazilian carnival queen.) Having dispensed with his Victorian obligation to aesthetic sensitivity, Salvin quickly fulfilled his enduring, if not primary, motive. 'And a moment afterwards, it is in my hand – the first Quetzal I have seen and shot.' Salvin shot his Quetzal with no more sense of guilt than a contemporary bird-lover feels when he sits down to a turkey dinner on Thanksgiving.

It took until the 1940s before a life history was written of *Pharomachrus mocinno*, the Resplendent Quetzal, and then only because of a truly unusual man, Dr Alexander Skutch. He roamed the Central American isthmus, studying its bird life for two decades, and is now considered an Audubon of the American tropics. In contrast to Salvin, Skutch decided early in his career never to kill a bird in the pursuit of knowledge. Nor to trap live birds, nor band them, nor interfere with them in any way. In fact, Skutch wouldn't even *look* at the stuffed Quetzals he passed in the homes and shops of Guatemala. 'This was not the way I wished to see Quetzals,' he once declared in an article. If it took years of patient, painstaking observation before the Quetzal revealed its manner of nesting and rearing young, which it did, Alexander Skutch humbly waited. His lonely work became an act of nearly spiritual investigation.

Skutch did not encounter his first Quetzal until more than two years after arriving in Guatemala, on a forested mountain called Cerro Putul in the northern part of the department of El Quiché. Even by the 1930s, hunting and extensive destruction of the Quetzal's habitat, as Skutch ironically put it, 'helped me to understand why it has taken me so long to see a living Quetzal'. Three decades had already passed since Guatemala had legislated protection for its national symbol, creating stiff fines for Quetzal-poaching of the kind Salvin encouraged. Yet although the law was apparently enforced, Skutch concluded that 'Quetzals must continue to owe their existence more to the inaccessibility of their haunts than to human laws, which . . . are usually not made until the creature they would save becomes rare almost to the vanishing point.'

For this reason, among others, Skutch never followed up his first glimpses of the Quetzal in Guatemala. He waited until he travelled to Costa Rica in 1937 before spending a full season observing a nesting of Quetzals. Then, as now, Costa Rican Quetzals were more abundant, in view of the greater areas of inaccessible subtropical forests, and a fairer land tenure system. In the past two decades, Costa Rica has established an extensive system of national parks and wildlife reserves, established with the financial resources the tiny country saved by abolishing its army in the 1948 revolution. Guatemala's Quetzals should be so lucky. The southern, or Costa Rican, subspecies of Quetzal differs from the northern, or Guatemalan, Quetzal primarily in its shorter, narrower tail plumes on the male's train. Otherwise *Pharomachrus mocinno mocinno* and *Pharomachrus mocinno costaricensis* are both much as Skutch described them in a 1938 journal, worth quoting at length for its thoroughness, accuracy, wealth of detail, and delightful prose:

The male is a supremely lovely bird; the most beautiful, all things considered, that I have ever seen. He owes his beauty to the intensity and arresting contrast of his coloration, the resplendent sheen and glitter of his plumage, the elegance of his ornamentation, the symmetry of his form, and the noble dignity of his carriage. His whole head and upper plumage, foreneck, and chest are an intense glittering green. His lower breast, belly, and under tail coverts are of the richest crimson . . . The dark, central feathers of the tail are entirely concealed by the greatly elongated upper tail coverts, which are golden green with blue or violet iridescence, and have loose, soft barbs. The two median and longest of these coverts are longer than the entire body of the bird, and extend far beyond the tip of the tail, which is of normal length. Loose and slender, they cross each other above the end of the tail, and thence diverging gradually, form a long, gracefully curving train which hangs below the bird while he perches upright on a branch and ripple gaily behind him as he flies. The outer tail feathers are pure white and contrast with the crimson belly when the bird is beheld in front, or as he flies overhead. To complete the splendour of his attire, reflections of blue and violet play over the glittering metallic plumage of back and head, when viewed in favourable light.

Much as I admire Alexander Skutch's oriental patience, the days when you could ramble into the Guatemalan highlands for a year are gone – or perhaps not yet returned. As a matter of fact, ornithologists have quit Guatemala altogether in recent years. Most archaeologists, too. Even the hippies who used to hit the gringo trail for the fabulous weavings and favourable exchange rate have bid good riddance. The political situation doesn't lend itself to timeless scholarship, much less joy-hogging. For one thing, the government will give you only a thirty-day visa. Yet it was something Alexander Skutch once wrote, in the sadly defunct ornithological journal *The Condor*, that helped inspire me to follow our course in Guatemala:

The Quetzal is something more than the living representative of a beautiful country of the present era; its human associations stretch back into antiquity. Possibly no other feathered being of this hemisphere . . . has a longer history, as the philologist rather than the naturalist would use the term. This history is largely unwritten; and it is to be hoped that before long [some-]one . . . will make good the deficiency.

If we want to know Central America, someone ought to peek beyond the dictators and the dominoes, to the ways living things relate to their environment. What befalls birds as different as the Quetzal and the Zopilote reflect and foretell what happens to humans: in the short run, ecology is natural history; in the long run, it's more like prophecy. So despite the dangers of traipsing in zones of conflict, despite the decline of the Quetzal, despite the rainy season, and our notable failure to procure safe conducts, we decided to leave Guatemala City for the highlands. To be sure, there was still uncertainty about precisely where we were headed. But this was to be expected in a country where no guides to the flora and fauna are available; where the death squads sit in black-glassed ranch wagons at the traffic lights, and where the president explains his policies to visitors by handing out Bibles.

At least we would have plenty of maps.

II

Roadwork

I

The distance from Guatemala City to the highlands of El Quiché Department is about 175 kilometres as the vulture flies, a negligible trip along a nice, graded, four-lane superhighway. But in Central America nothing of the kind exists: a trip of 175 klicks could take the rest of your life. 'Give us good roads,' went an old Mayan prayer, 'straight, beautiful roads.' But only the part about beautiful was answered.

We spent the better part of the morning just trying to rent a *quatro*, or four-wheel-drive vehicle, from the agencies along the Reforma. They gave us dubious looks, as though undecided whether to laugh out loud or hide under the desk. 'You must be going very far,' said one creep. 'Foreigners don't usually go far in Guatemala.'

'If that's so, then why are there no *quatros* for rent in the capital?'

The creep shrugged: it was just another of those Central American imponderables. 'We're with the CIA,' said Mickey at another agency. Without blinking, the guy answered, 'I can get you a new Range Rover the day after tomorrow. Unlimited mileage.'

But we couldn't wait that long. In the end we settled for a spiffy, metallic gold Toyota four-wheel-drive wagon that would have looked more in place under the shade trees of some affluent state-side suburb: it would be like travelling in disguise. Only one paved road, part of the Pan-American Highway, headed out of the city west and north, on a curve that would take us through Antigua, Chimaltenango, Chichicastenango, Santa Cruz del Quiché, and finally eastward into mountainous cloud forests of the type the Quetzal has apparently inhabited since its origin as a species, no one knows how long ago. Miles of slums ringing the capital lay in our way. They were freshly sprouted jumbles of scrap-board shacks, small markets, and workshops, full of dust, soot, exhaust fumes, and human misery. There were no road signs, no map would help us. The streets had no names, the houses were unnumbered. The people we asked for directions either had never heard of Highway CA-1, had heard of it but didn't know where it was, knew where it was but couldn't explain how to reach it, could explain but were mistaken, gave us an answer merely out of politeness or because they were afraid, directed us to the Ministry of Tourism, pointed in several directions at once (groups), or fled inside and bolted their doors as soon as they saw the Toyota slowing down. A young woman stood in her doorway watching traffic flow by, a thin white cardigan sweater wrapped around her shoulders. I asked her, 'Do you know the road to Antigua?'

'*Saber*,' she said, which is the Guatemalan equivalent of 'who knows?'

'Well, what's the name of this street we're on here?'

'*Saber*,' she repeated.

'But you live here, on this street, no?'

'*Sí*,' she said, gazing around as if for witnesses.

'You don't know your own address?'

She hesitated, and then said tentatively, 'It's Guatemala, isn't it?'

A harbinger of things to come. We were completely lost in a maze of carpenters' shops with the finished furniture displayed in the street, auto garages mired in rusty parts and junked gas guzzlers, shoemakers' benches, bus stops belching diesel fumes, little one-storey whorehouses washed pink or painted in tiger stripes with tattered curtains over the doorways and names like Bar Big Boy and Bar Good Time. Around and around we went in this seemingly endless shantytown, where a dirt street of hovels ran uphill as far as the eye could see towards a horizon blotted in grey dust, and the sewage flowed back downhill in open runnels. Dust, grit, smoke, weeds, garbage, slops, marl, excrement, packing crates, naked kids with bloated bellies, drunks lying in the gutters, and looming over everything an enormous billboard showing the arched ass of some girl in designer jeans. How do poor people take it? Why don't they burn the whole thing down and start over again? What have they got to lose? Another Central American question with no answer.

After several hours of this we drew up alongside a banana depot, slugged down several Fanta orange sodas, and went in to buy provisions and ask directions. Huge stems of newly arrived green bananas were lined up along the sidewalk, freshly cut from the trees. Inside the

small warehouse, the bananas were mounded up neatly to the corrugated roof. There were bananas of every conceivable variety: dark
shiny green Gros Michels, reddish-umber cooking plantains, leopard-
skin 'finger' bananas three inches long. For fifty cents I hauled out a
stem of fifty pounds, a small stem, but enough to last a week. The
depot owner kept a green parrot for his amusement. The bird ran in
circles around the concrete floor, screaming its head off, then whisked
up to the owner's shoulder and regarded me with cold condescension.
The man said, 'The road to Antigua? You're on it, *amigo*. Cruise to
the bridge and you're on your way.'

'*Gracias, señor*. Until later.'

'*Que le vay bien* – that it may go well with you.' And he stroked his
parrot, as though for good luck.

A few miles farther on we reached the bridge the banana man had
mentioned. It was a solid bridge, steel girders over concrete pilings,
and it looked rather new and expensive. There was nothing under it –
no water, at any rate; just a skunky gully and a billboard that pro-
claimed, 'Another government project we paid for with taxes'.

On the other side of the bridge was parked a Chevy Blazer with
black smoked glass, surrounded by guys in sunglasses with all too
obvious bulges under their polyester sports jackets. We quickly
reviewed the rules of the road for travelling into the Central American
countryside.

Never travel after dark.

Always be courteous at roadblocks.

And never give any lip to men driving Chevy Blazers with black
glass.

Fortunately, they weren't stopping anyone. They slouched against
the hood of the Chevy like lizards lolling on a rock, watching the
insects crawl by. You had the sense that they might get hungry again
soon. You had the sense they couldn't control their appetite.

Finally we seemed to leave the city behind. The traffic thinned out,
the road rose rapidly, and the air thinned, too. I spat ten times out of
the window, and my head cleared of all the urban guff and filth.
Negotiating tight switchbacks blasted through steep rock, we began
our initiation into the dreamlike Guatemalan countryside. Along the
roadsides walked a constant stream of Indian women with their tur-
quoise-and-white plastic water jugs balanced on their heads, or baskets

of fruit, or loads of firewood, balanced by tumplines and carried down from where their men had cleared the trees for planting maize. They wore their dazzling and justly famous native costumes, band upon swirling band of purple, yellow, red, blue. Their dress consisted of a skirt, a white *huipil*, or blouse, embroidered around the neck, then a shawl around the shoulders, thinner woven straps to carry their babies around their waists, and *puys* – woven hairpieces. These elaborate wrappings are modest and completely hide the feminine figure, yet aren't Victorian in the sense that they don't constrict movement. On the contrary, away from the confines of the city, the women seemed freer, easier, more animated in their movements. Their eyes stared straight ahead, but their arms and hips swung loosely in a dynamic and appealing rhythm. It was a gait that could only have developed among surefooted people confident of a high standard of roads long before the pavers came through laying the Pan-American Highway twenty years ago. In truth, the Indians built their goods roads hundreds of years before the Spanish Conquest. In his book *The Rise and Fall of Mayan Civilization* the archaeologist J. Eric S. Thompson compared the Mayans to the Romans in their finished roadwork:

In the days of their greatness, macadamized roads, raised from six to eight feet above the ordinary level of the country and surfaced with hard, smooth cement, led from palace terrace to temple, from temple centre to temple centre. Such highways radiated from Chichén Itzá toward all the other great centres of population . . . The Mayas a thousand years ago built their highways on practically the same principle as McAdam adopted in the nineteenth century. The roads, from twenty to thirty feet wide, were beautifully ballasted and bound in with stones. The old road-builders had found out that, by putting the stones edgewise and cementing them together, they would not readily dislodge. The rains served merely to weld the mass together and there was no frost to act as a disintegrating agent . . . Smooth these roads had to be, for the traffic over them was all of barefoot or sandalled Indians. It is apparent that the ancient Mayas did not possess a single kind of useful domestic animal or beast of burden, and the principle of the wheel was unknown to them.

Long ago abandoned, buried, split by tree roots, the Mayan roads remain only as archaeological curiosities, but that smooth, energetic pace of the still-barefoot Indian women has survived perfectly intact; they continue to walk everywhere, though on someone else's roads, not as well maintained as the roads of their ancestors.

The first *milpas* appeared, tiny to mid-sized patches of maize planted right into the karst hills rising above and falling below the road. They looked nothing like the American cornbelt, with its regular rows, its Anglo linearity. The plants were jammed chock-a-block, like the amber waves of grain North Americans sing about, except these were green waves of corn, green corn everywhere. The rises were steeper than it appeared possible to plant, yet they were meticulously worked, and now, in July, the maize was waist high and just beginning to tassel. Here and there we caught sight of a *campesino* peeking out of his green maize: the brown dot of his face and the glint of his machete. The corn tassels peeked out, too, from under clouds that drifted apart and re-formed around the humps of the hillocks like shy, scampering children.

We began to experience a giddy sense of soaring, produced not only by the precipices and the swaying road, the billowing clouds and the stilettos of sunlight stabbing at the lush green valleys below, but also by the companionship of the Zopilotes. The vultures patrolled the highway like airborne security forces, three or four of them floating deftly over every pass. The birds rode low to the surface, smart bombs on a hot wind, their black wings splayed in a wide V shape in order to readjust their course beside the harshly striated ravines. Unrestricted by habitat, they slipped into the thermals rising up along the mountain ridges, and soared along for hours as if on moving sidewalks. They flapped their wings once in a while, but just for something to do.

The roads are their dining rooms. They hang about waiting for road kills – the occasional horse or pig that strays into traffic, or the cow that lies down to a stubborn bovine death. From the looks of the way the drivers acquitted themselves, keeping to the roads must earn the Z-birds a good living. The buses and trucks wound their gears on the uphill grades, forming peeved convoys. As soon as the slower vehicles made for the rise, everyone behind them raced for the left-hand lane to get past, leaning on their horns, oblivious of the blind curves. If they didn't let oncoming trucks and buses slow them down, they weren't about to let a little thing like a loose domestic animal stop them. That the paved road suddenly became in places a dirt road, never announced by any sign, sometimes on a hairpin curve, must also, at times, give the vultures a taste of a higher grade of protein: car accidents were commemorated with a cairn of rocks and a crude cross, on-site burial being the custom for traffic fatalities. But that day we could only surmise the vulture's diet and method, not observe it.

2

Antigua was a pretty town – a town of pretty accomplished swindlers, pretty good con artists. They were waiting in a dense pack just over the small concrete bridge, the only road into town. There was no way to avoid the ambush. Gold teeth flashed beneath dark glasses, strings of postcards fell open like fake handshakes, trinkets danced in front of the windshield, and arms shot out holding government IDs that read 'Official tourist guide'. Antigua was once a favourite oasis on the gringo trail, but that was fifteen years ago, when Americans still visited Central America for its archaeology and landscape. Now Antigua was a ghost town and a ruin. The golden Toyota four-wheel was probably the best piece of work Antigua had seen for a while, but not much business to divvy up among a squadron of hustlers.

Our guide in Antigua was Sergio, a sullen, swarthy Ladino who won the job by accomplishing a daring mobile transfer from his old black bicycle into the back seat of the Toyota. Before we had time to figure out how he had accomplished this – he had thrown himself bodily through the window, leaving his wheel spinning in the middle of the road – he sat up with a smirk of pure greed and passed forward his credentials, assuring us he charged only the official rate of ten quetzals for the day. He was already negotiating our itinerary of sin.

'You want to change money?'

'No.'

'You want a woman?'

'No.'

'Marijuana? Cocaine?'

'No.'

'Archaeological treasures? I know the cheapest place to buy them.'

'Is this guy for real?' said Mickey in English.

Sergio must have understood some English, because he immediately put his ID back into circulation to confirm his reality. He pointed to his snapshot, to his name, to the words '*Guia oficial*'.

'Don't worry,' he remonstrated, 'I've taken a government course in tourism. You could say I work for the government – *official* guide. I know Antigua like I know my hand.'

He showed us his hand. He reversed it and showed us the back of his hand. I began to think Sergio had hands he wasn't showing.

'You want disco tapes?'

'No.'

'How about a pair of nice *huaraches*? Only five quetzals.'

'No, thanks.'

'A hotel? We have hotels with hot water, or with cold water only, seven quetzals. We have them with private bath, more expensive, and with –'

'Look, *amigo*, aren't there any sights here?' I interrupted.

'Sights?' he turned pensive. 'Oh, certainly, so many sights. Antigua is one great historical sight, the whole town. I'll show you – only ten quetzals. Antigua was the capital of the United States of Central America. Our Washington! – you're North Americans, right?'

'German,' said Mick.

'Ahh,' said Sergio. 'That's in Europe, no? Ha! See? I can guide you best. Ten quetzals, only.'

'Only?'

'Only,' he pledged with his hand over his heart.

'OK, Sergio. You're on.'

'Turn here! Cruise this street! More there!'

We drove around town while Sergio named all the churches, great Spanish Gothic horrors resting in the fallen stones of their own collapse. The old capital spread over a pleasant green valley of coffee plantations and fairly new suburban developments, all covered by a faint yellow volcanic haze. The twin volcanoes Agua and Fuego rose like smoking cannons from the ends of the south streets, with the other volcano, Acatenango, off to the west. It was like being surrounded by leering bullies. There was hardly a building that hadn't suffered some sort of earthquake, volcano, or flood damage – a jagged crack running up the façade; a corner separated from, and four feet below, the rest of the building; fallen masonry lying in a heap before carved and studded colonial doors; charred walls and adobe eaten away by fire, water, ash, and fungus. Sergio said the whole town was under renovation by government edict – man fights back – but it reminded me of the Mexican folk tale of the railroad company that decided to build a bridge through the valley where the devil lived. Each day the company spanned the valley with greater and greater girders, stronger and stronger ties, while the devil took his siesta. And each night the devil amused himself by demolishing everything all over again . . .

On every street of Antigua lay the ruins of another church, convent, monastery, bishop's palace, university, or colonial residence. It was difficult to grasp what would make people continue to live at the foot of three of the most volatile volcanoes on earth, in a place that has perhaps been visited by more calamities than any other city in human history.

Antigua was the second capital of the Spanish kingdom of Guatemala, founded in 1542 after the first capital, only a few kilometres away, was destroyed by a tremendous explosion of the Agua volcano. That event took place on the night of the very day the Spanish conquistador Pedro Alvarado's twenty-one-year-old widow, Doña Beatriz, angered church authorities by proclaiming herself ruler of Guatemala in her deceased husband's stead. A woman in charge? The priests predicted something terrible would happen, and their augury came to pass within twenty-four hours: total devastation. They chalked it up to God's wrath, moved to a new site, and started over. Nevertheless, Antigua's annals are no less an unending disaster saga, despite having for ever after been ruled by males. A quick browse through Antigua's chronicles tells the tale:

In 1558 an epidemic disorder, attended with a violent bleeding at the nose, swept away great numbers of people; nor could the faculty devise any method to arrest the progress of the distemper. Many severe shocks of earthquakes were felt at different periods; the one in 1565 seriously damaged many of the principal buildings; those of 1575, 76 and 77, were no less ruinous . . .

The years 1585 and 6 were dreadful in the extreme. On 16 January of the former earthquakes were felt, and they continued through that and the following year so frequently, that not an interval of eight days elapsed during the whole period without a shock more or less violent. Fire issued incessantly, for months together, from the mountain, and greatly increased the general consternation . . .

On 18 February 1671, about one o'clock, afternoon, a most extraordinary subterranean noise was heard, and immediately followed by three violent shocks . . . which threw down many buildings and damaged others; the tiles from the roofs of the houses were dispersed in all directions, like light straws by a gust of wind; the bells of the churches were rung by the vibrations; masses of rock were detached from the mountains; and even the wild beasts were so terrified that, losing their natural instinct, they quit their retreats, and sought shelter from the habitations of men . . .

The year 1686 brought with it another dreadful epidemic, which in three months swept away a tenth part of the inhabitants . . .

The year 1773 was the most melancholy epoch in the annals of this metropolis . . . About four o'clock, on the afternoon of 29 July, a tremendous vibration was felt, and shortly after began the dreadful convulsion that decided the fate of the unfortunate city.

I was not surprised later to find out that a specialized literary form evolved in Antigua known as the 'Earthquake Account'. Sometimes these were merely factual reports. More often they were embellished into allegoric legends of doom at the hands of infernal forces. The authors often repeated the old Mayan belief that giants slept inside the volcanoes, and every so often turned over in their sleep. It didn't cheer us up much to learn from Sergio that a team of North American geologists was in Antigua even at that moment, taking seismic soundings to see if they could predict when the next instalment of the catastrophe would begin: there had been rumblings for four weeks.

The Palace of the Captains General bore down on the Plaza de Armas like a cranky inquisitor. Its sombre double tier of arches, resurrected from the disaster of 1773, was immediately recognizable from the famous drawing Frederick Catherwood made when he came to Antigua with the explorer John Stephens in 1840. By then Antigua had become the nominal capital of the short-lived Central American Federation, before Central America split up into six separate turfs, six little gangs plotting against each other, six oligarchies spreading their venom, six little economies trying to grow the same things and export them to the same markets, and, of course, four standing armies (Costa Rica abolished its army in the revolution of 1948; Belize remained a British colony until 1981). No, the United States of Central America was an idea whose time came and went between eruptions here in Antigua, where the fractious, powerless, and completely penniless constituent assembly met a few times in the Palace of the Captains General, which has now been turned into an army garrison. The soldiers were standing guard on the veranda under those gloomy stone arches, but what was there left in Antigua worth guarding? A nation that never happened? Historical rubble? It was hard to say.

There were perhaps a dozen soldiers, and I doubt a razor had ever nicked any of their faces. They had stylized their camouflage fatigues so that the trousers clung smartly against their calves, thighs and crotches, and they kept their smokes rolled up in their meagre biceps. They all wore the darkest shade of sunglasses. A few had on camouflage berets. The effect was of a teenage gang cast for a remake of *Torrid Zone*. But they did have a .50-calibre machine-gun trained on the plaza, and they were carrying German G-3 assault rifles – expensive, state-of-the-art stuff that anyone can kill with after five minutes' training. So we decided to avoid them.

Instead, we ate bad chop suey in a filthy restaurant while Sergio tried to arrange a money exchange with the Chinese man who owned the place. The rate in Antigua was far below what we'd got from the Boy Wonder in Guatemala City, so we paid our bill and left. There were only a couple of hours of daylight remaining, so I told Sergio that what we really preferred was to drive out towards the volcanoes and see a bit of the countryside.

'I know just the place.' He rubbed his hands together. 'Take this street. Cruise!'

Mickey pointed the Toyota towards the green cleavage of the volcanoes, but well before we could reach their base Sergio directed him to skirt around through a coffee plantation. There were white adobe walls on either side of the dirt road, and 'Keep out' signs all over the place. Sergio said, 'It belongs to the Herrera family.'

He didn't have to say more, because I knew the Herreras were not only one of the wealthiest Guatemalan families, but one of Guatemala's infamous fourteen families. The fourteen families were reputed to own Guatemala, or at least the parts of it worth owning.

Prime farmlands, banks, factories, fertilizer concerns, and until very recently, the military and the goon squads, too. You definitely didn't want to cross any of the fourteen families, though you might want to marry one of their scions. But that's impossible, because their sons and daughters marry each other. They didn't get to be the fourteen families by opening their hearts to humble outsiders. In fact, the way they got to be the fourteen families was by resolutely denying that they existed as the last true oligarchy in the Americas, and by using plenty of muscle in their business dealings. At one time the fourteen families maintained private armies as enforcers on their haciendas and plantations, to repress rebellions by the peons. Then they decided it was cheaper to have the national military do the job. What they didn't figure on was that after a while, even the dullard military officers would get ambitious, and want a cut of the action.

Nowadays it's a real question as to who owns Guatemala – the fourteen families, or the generals with their big homes, big cars, big heads. Audacious generals have even been known to expropriate properties belonging to the families for themselves. This is called land reform in Guatemala. Members of the fourteen families are said to have been associated with the formation of the death squads, a Guatemalan creation. Other members have beaten a retreat to their condominiums in Miami, to count their money in the sunshine. Not the Herreras, by the looks of it. Behind their white adobe walls the silky green coffee bushes ran on for several kilometres. They were shaded by banana trees. The reddish black volcanic soil sent up a rich smell of humus.

'Do any of the Herreras live in Antigua?' I asked Sergio.

'No. I don't know. It could be,' he glowered. 'Why do you ask so many questions?'

At the far end of the farm we started down a steep dirt road, which became lined with Indian dwellings, oddly open-faced to the road. In almost every one an Indian woman sat at a wooden hand loom. Children ran into the road, shouting, '*Pase adelante . . . precios baratos!*' Which means 'Come in . . . cheap prices!' A few even ran alongside the Toyota like little dogs, trying to get us to stop and come to their mothers' shops.

'And what's this, Sergio?'

'Artisans.'

Mickey cursed under his breath. 'Now he's going to take us shopping.' But his eyes had already begun to light on the weavings displayed on the walls of the shops. And then all either of us could do was gasp, 'God, they're beautiful.'

It was a village called San Antonio Aguas Calientes, a strange name that could connote either the presence of a hot spring, or Saint Anthony as the patron saint, or maybe both, in the form of saintly protection for the villagers from the volcano hovering right over the place. We reached the dusty square. There wasn't a single tree, flower, bench, or fountain, just a vacant space with hard-packed dirt and two standing buildings. One was an old adobe church, seventeenth or eighteenth century by the looks of it, pretty smacked up and tumbled down from the natural abuse. The other was also a church, one of those charming small village churches you see so many places in Central American villages, elegant in its humility, a single arch on the façade, with a black iron bell rung by pulling a rope out front. It was painted sandy yellow with blue trim, and it was closed. The rest of San Antonio Aguas Calientes consisted of a hundred or more open-faced weaving shops, fronting on all four sides of the plaza and running back in several rows. The village didn't seem to have any men. It also didn't have any customers. If business was bust in Antigua, you could only imagine how bad it was in San Antonio Aguas Calientes.

As soon as the dust settled around the Toyota, you could see the eyes of all the women blaze up at once. A mob of barefooted kids besieged the car. We were water arriving in the midst of a long drought, food in the midst of great famine. There was absolutely nothing we could do about it. Sergio had us in the palm of his hand, maybe the one he hadn't been showing.

We drank Fanta orange and toured the shops. You could hardly remain disgruntled examining such weavings. Each one was more inventive and colourful than the next, though they were all equally beautiful, like trees in autumn when the leaves change. Each time we prepared to leave one shop for another, prices were slashed.

'What do you offer for this?' said the women. 'Make me an offer.'

'Gee,' said Mickey, 'looks like we came during the annual sidewalk sale.'

'You like that one? You like that one? Only fifteen quetzals,' Sergio kept saying. 'You'd pay twice as much for that in Guatemala,' he said, as though we had somehow left the country. (I later learned that Guatemalans commonly refer to Guatemala City only as 'Guatemala', the way Mexicans refer to Mexico City as 'Mexico'; belatedly I forgave the wretched woman in the white cardigan sweater who had tried to tell us we were in 'Guatemala'.)

What was remarkable was how many of the textiles included Quetzals or other birds in their motif. Not all the Quetzals were the conventional green with bright red bellies and red eyes. There were orange, red, and yellow Quetzals, too. A few Quetzals even had tremendous spread wings. They looked suspiciously like firebirds, or eagles. Interestingly, none of the Quetzals was portrayed entwined with serpents, as in the familiar Mexican national symbol, which seems to symbolize an eternal struggle between heaven and earth, love and death. The Mayans of today, like their ancestors, tend to put opposite and hostile characteristics into one deity. I found a little girl, well-spoken in Spanish, and asked her why there were so many bird designs, and if she knew what they meant.

'It's because of the patterns,' she said most seriously. 'We copy the patterns from books.'

'What books? Where do the books come from?'

'Various places. The book my mother uses she got at a hotel in Antigua, where she was working as a maid, but the hotel closed. They use this book in the factories in Quezaltenango, where the machines make the weavings. But we make them by hand here in San Antonio. The colours are stronger, and the weaving is better, and also, the prices are cheaper. How much will you give for this one?'

'About the book,' I continued. 'Can you show it to us?'

'I would let you see the book,' the girl said. 'But my mother's not here right now.'

'Then the patterns – they're not traditional patterns?'

'Traditional. Yes, they are traditional,' she insisted firmly, 'but we take them from books.'

You couldn't argue with that. I paid her for the weaving. We bought several more textiles with Quetzals in the next shop and in the one after that, some weavings without the bird, paying the prices asked – I wasn't in the mood for any hard dealing with women and children, and the prices were pretty low anyway. But when we returned to the Toyota to leave, I noticed Sergio wearing a hangdog face.

'What's the problem, Sergio? Didn't we buy enough?'

'Oh, it's not that. It's that you didn't appreciate the handicrafts.'

'But we did, very much. We bought a lot. Didn't you see? You were there.'

'You didn't make an offer. The artisans don't think you appreciate their weavings if you just say, "That's nice," and pay what they say.'

It was a form of praise or flattery we weren't used to, coming from a culture where to haggle over the price of something is an insult, into one where the insult is not to haggle.

Up the hill, on the way out of San Antonio, the kids all ran out on the road again, and this time, knowing it was their last chance, they were screaming at us, '*Más barato! Más barato!* – cheaper! cheaper!' So we pulled up again and bought a few more of the fantastical weavings of San Antonio. I counter-offered a little to placate Sergio.

By now, however, we'd had enough of Sergio, so when we got back into Antigua I paid him fifteen quetzals – the ten we owed him, and five to get rid of him. We dropped him off where he'd left his black bicycle in the middle of the road. It was still there. He said he was available in the morning to continue our tour of Antigua, but I could claim in all honesty that we had no intention of staying another day. At the risk of having him turn up on the doorstep at dawn, we took his recommendation and drove down to one of the hot-water hotels across from the market-place. It was called El Castillo – The Castle. The place looked about as much like a castle as a camel looks like Whirlaway. But I figured, hell, you can bear anything for one night. I was wrong.

3

The Castle was clean enough; that wasn't the problem, though there was no hot water as billed, only a coin-operated gas jet attached to the bathroom wall that swallowed centavos and dribbled back a blue flame too feeble to heat the water. The problem was the desk clerk. He came to our rooms several times to see about the hot water, thumped the heater with his fist, and told us to wait – just a little while more, just a few more minutes. He was a thin, nervous little guy in pointy black boots and a tight-fitting nylon racer's jacket zipped up to his throat. He also wore tinted flier's shades, the kind worn by guys who go to porno movies. His name was Cornelio, and since his brother was an army draftee, Cornelio was in fat city: he didn't have to serve in the military while his brother was in service, he volunteered. And if his brother got killed, he wouldn't have to serve at all.

'Not that I mean for something to happen to my brother,' he added, rubbing his chin slowly. 'But these days the country has many problems. The nuns are training the guerrillas.'

The third or fourth time he came to bang on the hot-water heater, he decided to wait around himself until the water got hot. And since it didn't, we could neither shower nor get rid of him. He lazed across the armchair and, in a manner that managed to combine machismo and cloying deference to Americans, started appraising our gear. He wanted to know how much everything cost, Mickey's cameras, tripod, my field-glasses, down to our three-dollar Japanese watches. All these gringo consumer goods set his teeth on edge with envy, made him smack his lips with impotent fantasies. There seemed to be a direct connection in his mind between gringo affluence and gringo sexual prowess, for even as he stared longingly at our gear, he moved right along to the subject of women, girls, little boys.

'I can get you anything you want, no problem,' Cornelio boasted. Finally, to get rid of him, I told him we could talk with him about plans for the evening later on. As soon as he was gone, we bolted the door, closed the shutters opening on to the first-floor courtyard, took cold showers, and hit the sack.

That was only the beginning of a night to remember, a confusing, scary, Central American sort of night. Within a few minutes, Mickey was sleeping like a contented baby. Good sleeper, Mickey. It was one of the things I'd noticed about my partner when we'd first met in Nicaragua. He didn't drink, he didn't smoke, and he could sleep through anything. Gunfire, explosions, traffic, bad weather, personal danger: whatever mayhem happened to be going on around him, he would be freshly bathed, hair neatly parted, stretched out on his bed or sleeping bag at seven or seven-thirty in the evening. Then he listened to the BBC World News or the Voice of America on his shortwave. By the time the news was over, he was soundly snoozing. What took place after the sun went down, and he could no longer take photographs, simply didn't interest him: the embodiment of Flaubert's recommendation for artists to 'live like a bourgeois and be wild in your art'. I was not surprised, when I got to know him better, to learn that his favourite photographic subject was Third World children warriors: thirteen-year-old Lebanese manning anti-aircraft batteries, seven-year-old Nicaraguans posing with their Kalyshnikovs. Yet he gave no indication that his self-discipline was difficult to impose on a contrary nature.

'It's my precision German engineering,' he'd quip, referring to his Wisconsin-German family. 'It's my Midwest *rutz*.'

I couldn't get to sleep so early, however – it was still no later than 7.30 or 8 p.m. I dressed and slipped out of the hotel, found a corner bar, and went in for a drink. The place was a little larger than a pantry, more the size of a walk-in closet. A yellow light bulb dangled on a single strand of wire above a chewed-up wooden counter. The counter was a little sideshow of tiny bags of plaintain chips, popcorn, caramel corn, cheese-flavoured titbits, sugar-coated peanuts, and a thousand varieties of horrible artificially coloured sweets. There were a couple of men slumped over Cokes at the only table, which was wedged into the corner right between the front and side shutters. But most of the clientele were filthy street kids, stopping in on their way home from work as shoeshine boys, vendors, pedlars, or beggars, for a cheap stand-up supper of junk food and Coke. At home their mothers would be waiting with the traditional meal of *tortillas* and beans.

The bar had nothing else to drink, except a few dusty bottles of unlabelled 'brandy'. I wasn't hard up enough to try the local moonshine, so I ordered my usual Fanta orange. While I stood sipping the soda, listening to the Central American pop music on the bar radio – a strange *mélange* of schmaltzy Mexican *mariachi* and dated American disco – Cornelio appeared at my side. I bought him a Fanta.

'Ready?' he asked.

'For what?' I answered – I'd already completely put him out of my mind.

'For a woman!' He was already excited at this prospect of playing the pimp, showing off his own masculinity by procuring for a foreigner, like a dog that gets aroused watching other dogs copulate.

'Sorry, not interested,' I told him. 'But I wouldn't mind a cold beer.'

'Same place,' he chuckled. 'I know a bar. It's in Chimaltenango.'

I don't know why I went with him. I wasn't the least bit interested in a woman, especially any woman Cornelio knew. I certainly wasn't interested in his draggy companionship. Antigua had struck me as a town that had died ages ago, and not just once, but repeatedly. Even after it was dead the first time the earthquakes and the volcanic eruptions, the plagues and the pestilences, kept right on hitting it, so the place had the quality of a stinking, mutilated corpse. You could feel an awful electric spark in the dead night air. You could smell the rigor mortis. It didn't do much for your sex drive. Probably, I went with Cornelio out of boredom, but a cold beer did sound enticing: the day's dust and filth had backed up in my throat to the point where it was hard to swallow. The soda sugar-coated it, but only made me more thirsty. It was definitely Miller Time.

'Is it far?' I asked Cornelio.

'No,' he said. 'We'll get a taxi. I have a friend with a taxi.'

'I don't want to know your friends. Let's walk.'

'You mean, by foot?'

'Is there another way? You said it's not far.'

'Oh, no, just outside of town,' he said falsely. '*Vamos!*'

The little hood didn't know what he was talking about or where he was going, but I was willing to endure that, if chasing the wild goose led to a cold brew. We walked for about a half hour without speaking, around the smouldering dump on the backwater of the market-place, past the ruins of a church where a family were living under a lean-to in the churchyard with their pigs and chickens, and then down a dark country road. Bats squealed overhead.

At length I said, 'O K, that's it, I'm turning back now. There's no bar down here. There's no nothing down here.'

'Yes, just a little more there, in Chimaltenango,' Cornelio insisted weakly.

We ended up in a graveyard, where I told Cornelio I wanted to look at the headstones, which were being visited in the still-early evening by people with lit candles, small bouquets of flowers, and other offerings. The marble and granite tombs stood in tight ranks. Some were old family sepulchres, built in the mysterious shapes of pyramids, obelisks, even ziggurats and pagodas. These were the mausoleums of the Spanish conquistadors and their descendants, the hacienda owners, the coffee barons, and Spaniards who had come to Guatemala under the sign of the cross, and perhaps got rich under the sign of the quetzal. There was a little boy playing among the tombs, and we struck up a conversation. He lived with his parents and eight brothers and sisters, he said, in a shack in the hills behind the cemetery. Two of his siblings had died the previous year, but weren't buried here for lack of money. The boy had been sick for six months with the same pneumonia that had killed them. His parents had thought he would die, too, but somehow his father had obtained medicine for him. He was recovered now, and going to school. But he spent most of his free time playing in the cemetery. He explained, 'It's nicer here than where I live. I like these buildings of the dead. It's like going all over the world. The dead get to live in better houses than the living.'

We turned and walked back to The Castle, and I left Cornelio in the lobby with a cold 'Buenas noches'. I was sitting up in bed writing in my journal when I felt the first tremor. It was as if a small demon had got underneath the bed, grasped the bedstead, and started rattling it furiously. The thought of having to call in an exorcist in the middle of the night scared me half to death. It took me a few more seconds to realize that the ground itself was shaking. But by then it had already stopped; I shut the light and went to sleep.

The rumbling started again about 1 a.m., and this time there could be no doubt about its cause. The shocks came in waves, lasting five to ten seconds each, and the ceiling shook and the pictures fell from the walls.

'What's going on,' said groggy Mickey.

'Get up. Get up,' Cornelio knocked on our door.

We rousted out into our clothes, and went out into the dark street to join the other guests – a German biker from Houston touring Central America on a BMW, and a nurse from Houston, no relation, as nervous as could be.

'Do you think the volcano will erupt?' she asked Cornelio.

'It is possible,' he snapped in an officious, pseudo-military way, as though awaiting further orders from higher-ups.

To allay the nurse's fears, I explained the geological theory that such waves of small tremors reduce tensions in the underlying strata, gradually releasing some of the energy that might otherwise build up and explode through the volcano. The tremors we were experiencing actually made a volcanic eruption less likely. Cornelio didn't understand English, and perhaps he thought I was putting him down. He gave the sky-blue muffler he'd added to his costume a toss around his neck, and went off in a huff to join a group of local men who had assembled on the kerb of the cobbled street. Under a distant streetlamp you could see that the cobbles had buckled like a roller coaster from all the tremors. But the street was mostly empty. Few Antiguans had bothered to leave their houses in anything like haste. One family came out, climbed into the back of their pick-up parked in front of their house, and promptly went back to sleep. Across the way, a woman leaned her huge bosoms out of the second-storey window and called down to see if an earthquake was expected. When no one answered her, she shut off the light and closed the shutters: the Antiguans were obviously accustomed to this sort of nightlife.

The men on the kerb, however, seemed to have settled in for a long vigil. They talked and smoked quietly. The trees fell silent, and the silence was stunning, as though the very molecules of the air had stopped moving. I didn't like the sound of that silence, the proverbial calm before the storm. I'd never heard of the phenomenon of supersilence described before an earthquake, but I had experienced it before hurricanes, and I'd already been through enough that evening to know that Central America doesn't always make sense. We stood in the street for forty-five minutes, waiting for Antigua to be destroyed again. Another jolt came, this one less severe. I was amazed at how quickly I was becoming sensitized to the severity of the shocks, like a human Richter scale. And how I had started to hope for the small friendly shocks. Maybe if we all just kept cool, the giants in the earth would stay asleep.

A woman came out of her house in a skirt and sandals with a sweater wrapped around her shoulders, went over to the men on the kerb, and announced sleepily, 'They're saying on the radio the volcano is sending up sparks and may explode any minute.'

'On the radio?' asked a gent in a poncho and hat, who seemed to be the elder statesman of the group of men.

'No,' said the woman, 'in Chimaltenango they're saying that, I heard it on the radio.'

'Is that so?' said the man. 'Well, then in Chimaltenango they don't know what they're talking about, because the volcano gives off sparks every month, yet here we are, standing in the streets of Antigua in the middle of the night.'

The woman accepted the rebuke with a yawn.

Cornelio said to the man in the poncho and hat, 'Then you don't think it's serious?'

Just then four creatures appeared at the end of the street, coming towards us with flashlights. As they drew closer, I could make out red crosses on their T-shirts. They were wearing gas masks, and holding strange meters in front of them as they marched past in high yellow rubber boots: it was how stories of interplanetary aliens get started.

'*Buenas noches*,' said the man in the poncho and hat.

'*Buenas noches*,' gurgled the crew through their gas masks.

When they had passed, the man in the poncho and hat turned to the woman and said, 'You see? Go back to bed.'

We went back to bed, with the addition of the frightened nurse from Houston, who was too shaken up to sleep alone on the second floor. Cornelio quickly offered to stay with her, but she turned him down flat, and asked if she could stay in our room instead. I agreed immediately, with some sense of accomplished revenge against the little creep. But no sooner had we arranged ourselves and dropped into an exhausted doze than the sharpest shock of all struck. This time, the whole building swayed dizzily. When I opened my eyes I saw Mickey and the poor nurse from Houston catapulting out of bed and colliding in mid-air. I'm certain no one in Antigua slept much after that. As for us, we lay back down fitfully, trying not to think about the things one thinks about in such circumstances, then finally went off together after first light to eat chocolate layer cake for breakfast. It was the only thing we could find at that hour in Antigua.

4

The road from Antigua led north up a twisting valley to Chimaltenango, a dusty assemblage of roadhouses, garages, and eating places, strung out along both sides of the road. People and animals stood in the road waiting, but for what no one could say. Others were sitting waiting in the doorways. They looked like they had been waiting a long time. Chimaltenango was ringed with lush, irrigated plains, ploughed and terraced in green maize and wheat, intercropped with beans, squashes, and peppers. The fields were neatly hedged by rows of tall magueys, the grey-green cactus with clusters of long, spiked, sword-like leaves. Above the gritty haze on the horizon towered wooded ridges masked in blue, marking the beginning of the highlands.

But even as we reached the foothills we came on a couple of dozen Zopilotes, busily devouring a dog carcass in the road. We stopped to watch, knowing that the vultures wouldn't mind the intrusion. Indeed, they paid no attention at all. Once they are on carrion, they are not only single-minded, but as of one mind. Trucks sped past, vans swerved at them for fun. Cars sounded their horns. The flock jumped in unison out of the way. Before the vehicle had disappeared they had all jumped back again and aggressively resumed their meal. You could approach within a few feet, and if you stood quietly, not making any startling movements, it was like being invisible. At such short range, you could begin to differentiate them – the slightly larger males with their flat, black, wrinkled face masks, the smaller females with their mousey grey sheathings. A few, maybe females or young birds, had masks with a bronzish shine. But the most marked difference was in age, not sex: the old vultures, with their worn, dry primary feathers, broken wing tips, bald patches. They must live to a stunning old age, but how old no one knows.

They had no true pecking order, but lunged at each other willy-nilly, snapping their beaks and flapping their wings like fighting cocks. They barked, woofed, snuffled, snorted, wheezed, and hissed at each other in a constant struggle for a place at the dinner table. Looked like one big happy family. Vultures displaced by other vultures hopped away, wings spread, shoulders hunched in irritation, only to turn around and try to drive off some other member of the flock – usually not the one that had driven them off in the first place. They all seemed to get enough to eat this way. They certainly did each other no harm in these flare-ups, which are nothing more than mock fights, for we never saw a single peck land, nor a single feather fly. Nor did they use their claws in these feints. All the flaring, jostling, flapping, seemed a highly formalized dance, like a professional wrestling match – a staged expression of aggression in a species where the flocking instinct won out long ago. This gregariousness, too, may account for why the vulture is voiceless, and does not possess more than rudimentary vocal cords. Birds' voices have developed primarily as a means for the individual to lay claim to territory and attract a mate. But vultures, practically alone among birds classified as predators, have given up, if indeed they ever developed, territoriality. A creature so singularly adapted to scavenging would need to compete for, and defend, an immense area. Instead, the vulture ranges rapidly on its powerful pinions over a wide area that it makes no attempt to maintain against others of its species. Though they don't actually reside in permanent flocks, vultures are social scavengers. Similarly, mating rituals in the vulture family have undergone a certain evolutionary socialization. Here is how Audubon described the Black Vulture's mating customs in *The Birds of America*:

At the commencement of the love season, which is at about the beginning of February, the gesticulation and parade of the males are extremely ludicrous. They first strut somewhat in the manner of the turkey cock, then open their wings, and as they approach the female, lower their head, its wrinkled skin becoming loosened so as entirely to cover the bill, and emit a puffing sound, which is by no means musical. When these actions have been repeated five or six times, and the conjugal compact sealed, the 'happy pair' fly off and remain together until their young come abroad.

This elaborate collective display has evolved as a substitute for the intense territoriality and rivalry of most other male birds of prey in mating season. And in the sense that communal ritual takes the place of individual competition, the social life of the vulture can be said to be fairly advanced on the evolutionary scale. One has only to look at the advantages accruing to a species whose members cooperate. For it would be unimaginable that even the fiercest bird of prey, working alone, could control the meat of a large dog, a horse, or even a human carcass – things these informal flocks of Zopilotes handle routinely every day in Central America.

It was difficult to say how fresh the dog carcass was, as it had been dragged around in the dust a good deal. Its smell might have given up a clue, but the prevalent odour was the sweet-sour stench of the birds themselves, which they develop early in life as nestlings and keep with them into adulthood, probably as a defensive mechanism, which signals other potential predators that vulture meat is tainted, and not worth attacking. This defensive, or, you might say, offensive, stench, is deemed so effective that practically alone among birds the vulture has no enemies. Not an eagle, not a man, not even a serpent will hunt the vulture; a remarkable achievement in nature, when you think of it.

We had arrived in good time to observe the methodical, almost mechanical, order in which the vultures made their meal. First they used the slightly hooked upper edge of their bills to poke a hole in the dead animal's abdomen, just above the anus. When the perforation had reached sufficient diameter, about three to four inches, several of the birds at once buried their long beaks inside, up to the nostrils, which sit very high on the beak under the vulture's eye. Seizing the viscera, they began tugging, until the guts slithered out in a long sausage. Meanwhile, the other vultures continued ripping and widening the cavity on the underside of the carcass, until all the organs lay exposed. The prized intestines and stomach, bladder, liver, and so forth were separated from the carcass and gulped down a short distance off.

Several other birds pecked at the eyes, till nothing remained but barren sockets. The vultures then crowded round the torso, each one clamping a clawed foot inside for leverage as they tore chunks off with their beaks. Their balance seemed precarious with one large foot planted inside the carcass, which shifted constantly as they all dragged, lunged, and tore, and the other foot as sole support on the ground. Were there right-footed vultures and left-footed vultures? I couldn't discern any pattern: they seemed to employ whichever foot provided the most stability at a given moment. From close range, I could clearly see that their grainy, nearly scaly, shins, lacking tarsal feathering, were covered in a white, limy scree. The theory is that the vulture purposefully defecates on its own legs. The highly acidic content of this whitewash acts as a protective coating against the germs and vermin encountered when the bird thrusts its feet deep inside rotting carrion and faecal matter. The vulture's stomach, too, must contain potent acids, enzymes, disinfectants, to render the vulture's rotten food harmless. And perhaps so much of the vulture's evolutionary energy has been spent developing these protective juices that it had nothing left for pretty colours, fine voice, or nest construction.

Only after the entrails were entirely consumed did the vultures turn their attention to the muscle tissue, but whether because it keeps better in the tropical heat, or is considered less delectable, I don't know. They tore the hide away, and while several of the flock engaged in a tug-of-war over it, the rest worried the meat off the bones. In forty minutes, the animal was reduced to a skeleton, save for the head, which was still intact above the neck, skin and all. Except for the eyes, the vultures ignored the head, as if mocking the custom of hunting predators, hawks and owls, which usually consume their victim's head first.

After a while, only two birds remained on what was left of the dog carcass, which wasn't much. Another half-dozen vultures moved clumsily down the road, finishing off various bits or trying in vain to lift their great, gorged bodies into the air. Stomachs and crops swollen, none had the strength to fly away. Instead, they fell into poses of utter sluggishness: time for a long siesta. Several hopped up on fence posts, or a low hillock near by, where they intermittently preened and roused their lower neck feathers, trying to clean off the blood and guts. Most of the flock, however, arrayed themselves across the roadway, and fell promptly into a deep torpor with their wings outstretched to let the heat and sunlight kill off the germs they'd picked up feeding. They looked like a roadblock, just another of death's checkpoints, and down the road came some barefoot kids carrying their schoolbooks. Suddenly, the children stopped short. They took one look at the torpid birds of death, another at the two white men standing over the skeleton, and without further ado, ducked into the nearby *milpa*. We could hear them scattering crazily through the maize, snapping off leaves and crushing stalks in their panicky flight to escape.

5

As soon as we got by Lake Atitlán and up into the hills, the country got dry and desolate fast. We had entered the Central American badlands, an arid stretch that was once confined to southern Mexico's Chiapas province, but with human assistance in the form of wanton timbering, unregulated livestock grazing, and the total lack of erosion control, has now formed a creeping desert through Guatemala. The land was blistered and burnt out, most of the trees cut down, the heat so intense that few crops were growing any more. All we saw were signs of drought and starvation: hungry peasants and grim soldiers, mud huts collapsing under red tile roofs, dead horses and dead Indians bloating on the roadsides, junked cars and razed farmhouses, vultures, hollow-faced children, and funerals in the road. The elevation was high enough for Quetzals, but there weren't any Quetzals here any more, and there would never be Quetzals here again. Every twenty kilometres was a military installation, where several acres had been levelled by bulldozer, turned into a minefield, and surrounded by high barbed-wire electric fences. Then for the next five kilometres there'd be wooden guard towers jutting up from this moonscape, with machine-gun muzzles sticking out of sand-bag nests pointed down at the road. The army was dug in like an occupying force, their main work consisting of keeping the roadsides clear of vegetation so that no one could get close enough to fire a shot at a guard tower, or ambush a convoy. This in itself was a daunting task, for the so-called 'pioneer' or 'volunteer' plants, the first ones to take root in patches exposed to sunlight wherever primary forest is cut down, are mainly tenacious vines, and quick-growing, large-leafed shrubs like cecropia and oak, all extremely tough to get rid of – as tough as putting down a peasant rebellion in a place where there's no food, no land, no farm work, no hope. When I saw how the army was going about its job of suppressing the jungle – with bulldozers and flame-throwers; pressing ragged Indians into road gangs to chop the vegetation down with machetes; spraying herbicides from back-packs and helicopters, I thought of something I'd heard the general say in a Sunday-night TV sermon I'd caught back in Guatemala City: 'Guatemala requires a change. That change will consist of one adversary imposing its will on the other.' What the general hadn't said was that the 'other' included not only the guerrillas and the peasants, but the plant kingdom as well. I wonder if the Bible-thumping general had ever read Isaiah's warning: 'Your country is desolate, your cities are burned with fire: your land, strangers devour it in your presence, and it is desolate, as overthrown by strangers.'

A light sprinkle of rain started, and the road grew greasy slick. The Toyota four-wheel held the course perfectly, but on one of the seemingly constant hairpin turns a flat-bed truck in front of us lost its load. It was a huge metal vat about eight feet in diameter, bound for a brewery or maybe a concrete yard. The thing shifted centrifugally as the flat-bed took the corner too fast. The momentum snapped the heavy iron chains holding down the vat like toys, and the vat slid right off. It narrowly missed clipping the Toyota across the hood, landed on the road, then skidded and bounced downhill, sending off sparks. The truck kept right on going: the driver didn't know he'd lost his cargo. We honked and gestured to get his attention, but when that didn't work, Mickey downshifted and pulled out alongside and we pulled the guy over. It was a pathetic scene. When he saw what had happened, the Indian truck driver broke down in tears. 'I'll lose my job for this, I'll lose my work,' he wailed. 'How will I feed my family? What will I do?'

Then the light sprinkle turned into a shattering downpour. The Indians walking along the road were busting their humps carrying their loads in the rain. Up ahead, an army helicopter had dropped a pile of corrugated metal roofing panels, and the Indians had converged on the pile like army ants, hoisting the panels on to their backs and scuttling away with them through the rain. The five-minute roof-replacement jobs were already appearing on the hovels and sheds for the next few klicks, the corrugated panels held down by rocks. It was the general's idea of a rural pacification programme: the same tattered, barefoot, hungry *campesinos*, standing under new tin panels dropped from the sky.

6

The same hard, dirty rain was falling in Chichicastenango when we got there, and an Indian funeral was going on in the streets. At the head of the procession, the wooden coffin was borne by six men. You couldn't tell whether they were in a frenzy of grief, or drunk, or both. They kept staggering and wailing, and dropping the coffin. The Quiché women were draped in dazzling rainbow mourning. Some were feeding their babies at their breasts even as they followed the pallbearers. Some of the older women were fainting. The men, in cheap trousers and torn shirts, unprotected from the rain even by plastic ponchos, were collapsing, weeping, rolling on the ground, beating their fists against the cobbles, and tearing their hair. Meanwhile children ran alongside the cortège, trying to sell sweets. The men carried the coffin up to the central square, a muddy cobbled market-place practically empty because it wasn't market day, and stopped in front of the famous El Calvario Church. It was here, in 1702, that the Dominican priest Francisco Ximénez worked on the Maya-Quiché mythology variously known as the Popol Vuh, Popol Buj, Book of the Community, the Sacred Book, or the National Book of the Quiché. The original document, whereabouts unknown, was apparently a compilation of ancient Mayan legends, set down shortly after the Spanish Conquest, during which the original Mayan texts were very probably destroyed. Not a written text *per se*, but more likely a book of paintings with hieroglyphs, which the native priests kept from one generation to another, and interpreted for the people. It contained the Mayan cosmic concepts, the origins and migrations of the Quiché peoples, accounts of their wars and alliances with other Meso-American tribes, a chronology of their kings down to the year 1550, and, perhaps above all else, the legends of Kukulcan or Quetzalcoatl, the Toltec deity whom the Mayans adopted and remained loyal to through all their rainy history.

No one is quite sure whether Padre Ximénez himself translated this aboriginal American bible into Spanish, or whether he worked with the local Quiché scribes around Chichicastenango as a kind of editor, advisor, godfather in the literal sense. What is known is that Francisco

Ximénez was an intensely scholarly man, who spent most of his two years in Chichicastenango studying the native language in the dank, porticoed monastery of El Calvario. He found the Mayan tongue so orderly, precise, and harmonious that he became convinced that 'this language is the principal one of the world'. Ximénez took a professional interest in the Indians' religious traditions and soon, by calling in the native scribes as his tutors, gained a confidence among the distrustful Indians that perhaps no Spaniard up to his time enjoyed. In the end, the Indians shared their legends with him. In his foreword to the Popol Vuh, Padre Ximénez noted the great secrecy with which the Quichés had previously hidden their lore, ceremonies – in fact their entire religion – from the conquerors. Yet, he wrote, 'I found that it was the doctrine which they first imbibed with their mother's milk, and that all of them knew it almost by heart.'

The Popol Vuh was one of the first evidences of the literary genius of the Mayans, and to this day remains perhaps the most useful evidence we have of how the Mayans experienced the world and recorded their collective life. More than two centuries later, the Guatemalan writer Miguel Angel Asturias, a Mayanist himself, would use the occasion of his Nobel Prize lecture to argue that the Popol Vuh was the first American novel, a richly imagined historical tapestry composed in utter secrecy by subjugated, oppressed people – and therefore, Asturias said, the forerunner of all of Latin America's magically realistic literature of struggle.

The pallbearers outside El Calvario started spinning the coffin around and around, faster and faster, until they grew dizzy and fell. The coffin crashed into the nearby mourners, sending half a dozen sprawling on to the hard, slick stones. We waited at the rear of the procession to get by. Now down the great sooted, cracked, hoary stone steps of the church hurried a pale, white-skinned priest, holding up the skirts of his long black cassock. He waded into the crowd like a true fisherman. From my vantage point on the outskirts, peering from inside the car through the inundation, I couldn't make out whether the priest hastily blessed the coffin, or was refusing the funeral admittance to the church. I couldn't imagine that it was the latter. Christianity has always been pretty liberal in Chichi, which is probably why so many Indians live in and around the town. By tradition, the Indians have been allowed to worship the Mayan gods in the plaza, where they hold their festivals, and the Christian deities inside El Calvario. But with time the two pantheons have blended, and today there's no difference; the native's religion and the conqueror's have completely fused. Whatever the priest had said or done, the coffin carriers did not ascend the steps. Then their mad grief seemed to increase. Several men flung themselves on to the coffin, ranting and raving and beating their fists. They had to be torn off and helped to the sidewalk by consoling relatives. The spasm of suffering spread through the crowd. It seemed as though all of Chichi was weeping. Where we were on the margin, men began pounding their fists on the car windows and kicking the bumpers wildly: they weren't threatening, just beside themselves with sorrow.

We decided to back slowly out of the procession and drive around town until the funeral moved on and the scene calmed down. There was a gloomy medieval chill in the dark little walled streets of Chichicastenango. The town's dominant colour was the smoky, grey-black of old volcanic stone and sorcerers. Only the heavy rain was keeping the vultures off the streets. We passed an Indian kid standing on a corner with a bucket of water who wanted to wash the car – an obscure, desperate desire, considering the downpour. He fell in behind the Toyota, and trotted along behind us. Whenever Mickey turned a corner, the kid turned, too. If Mickey hit the gas, the kid speeded up. There was no shaking him off. We ended up on a side street leading back up to the plaza, where we parked the kid and the Toyota and went into a small shop to drink a Fanta. There was another kid standing there, an Indian teenager in a loud green sports jacket, holding an oily red gas station rag in his hand. He watched us guzzle our Fantas in awe, followed us back to the car, and stood there gaping. God only knows what he wanted. Now we had two. As soon as we returned to the car, we were deluged by vendors like locusts, old Indian guys in dog-eared hats and wool watch caps trying to shove their souvenirs of Chichicastenango in our faces, little charms of metal painted gold, figurines of stone scraped to look old. One vendor pulled out a knife – I didn't know if he was trying to sell it to me or threaten me with it. We got into the car, rolled up the windows, and considered the situation. At this point we had one kid as permanent car chaser, a second kid standing there with an oily red rag in his hand, and several vendors draped over the windshield brandishing trinkets and daggers. To complete the cast, a shoeshine boy named Tomás now appeared rapping at the window, his spiky black hair shaved close as a muskrat pelt. Mickey gave him the go-ahead and stuck out his white Velcro sneakers, which Tomás fell to glopping with black shoe polish as though they were patent leather. The other Indians gathered round to watch.

'Don't buy any of their things,' Tomás murmured conspiratorially in good Spanish. 'They're not *real* things. I know where to get *real* things, cheap. I know what real things are worth.' With his flat-top haircut and round brown face, Tomás looked like a fallen angel, but his *spiel* betrayed him as a street-wise Central American kid hustler, the kind who hung out at the market-place and earned his daily *tortillas* by his wits. If you wanted a shoeshine, Tomás was your shoeshine boy. If you wanted a guide, he'd take you anywhere. And if, by some miracle, you wanted to buy something, he knew just where to find it in Chichicastenango, and what it should cost, with his take already factored in. He was only eight, or maybe nine, but he already had a creased, trapped, river rat look in his black eyes that said, 'I'm going to survive this sonofabitch world, one way or another, and you ain't going to stop me, mister.' You forgive things in an eight-year-old Indian kid you'd condemn in an adult like Cornelio. Perhaps it was only the evidence before our eyes that Tomás would never have the benefit of childhood innocence: he even bit the coin Mickey gave him to make sure it was real before slipping it into the pocket of his mud-coloured short pants, cinched to his malnourished waistline by a frayed elastic band.

This Tomás was one tough kid. He went at the other Indians, who were twice his size, with kicks, slaps, and curses, till they'd all skulked away. Then he led the way up the stone steps of El Calvario. It was a long climb to the top; you couldn't help but think of those pyramid temples of the pre-Columbian Mayans. The wooden front doors of the church were so gigantic that a smaller door had been cut into them where the Indians went in and out. A sign there said the church had been destroyed twice by earthquakes, once in the eighteenth century, and again in 1942. The sign must have been made before the great quake of 1976. Outside those great wooden doors stood a beggar burning pungent incense. He was a middle-aged man horribly disfigured by the Elephant Man's disease; his face looked as if it had

been put through a grater, then pumped up with air. I couldn't understand a word he said. 'He said,' said Tomás, 'he said, "It's a cold day
and may God bless you for the help you give him." Give something to
the ancient one,' he added in the same hushed and secretive way he
said everything. 'Give something to the ancient one.'

He collected a coin each from Mickey and me, slipped one into his
pocket, and placed the other in the beggar's hand. The incense sizzled
and smoked in the foggy rain.

Inside, El Calvario was dank and dark as a cavern. There was no
light but what came from the many candles burning on the small
altars around the perimeter. It had the rank smell of burning incense,
must, rotting wood, old stone, and melted wax. There were no pews
at all, only a bare floor laid of sepulchral hunks of grey-black stone.
The altar was so far away you could hardly see it. All along the floor
the Indians had laid bundles of herbs and flowers as offerings. The
place looked like a greengrocer's. Here and there in the dimness you
could make out the vague figure of an Indian, genuflecting or lying
prostrate on the small sheaf of straw he'd brought into the church as a
manger for his prayers. Above, the stained-glass windows on the left-
hand wall were all broken, every one. 'Earthquake,' observed Tomás.
'Last February fourth.' On the opposite side, the upper wall held
immense religious paintings, portraits of Spanish bishops and grandees
and things, framed in timeless gilt. But they were so faded and blackened that their subjects were hardly discernible: at a glance, they
looked like mysterious black renderings of eternal darkness. The only
other things my eyes could make out were the scaffolding at the front,
where some monochrome saints had been damaged in the recent
quake, and a sign asking visitors not to take pictures of the Indians.
'You'd need a strobe, anyway,' said Mickey, losing interest immediately. 'Let's go shopping with Tomás.' The oddest thing about El
Calvario, however, was this: it was so dismal it actually reinforced a
chilling sense of spiritual presence. The stony black silence, the dreary
wreckage, the smells of decay and entropy: they were like God's
meditation on this part of creation.

But I wanted to see the Dominican monastery attached to El Calvario, where Padre Ximénez had lived. Out the side door, the arcade
of the monastery had been racked by the recent quake. Several of the
supporting columns had fallen. A plaque commemorated Padre

Ximénez's work on 'the Mayan manuscript of Santo Tomás Chuilá', which was Chichicastenango's original name in 1540, when Central America's first refugees, Indians fleeing the Spanish Conquest, first settled around the Dominican outpost. 'Ah, *sí*,' said our own Saint Thomas when I read him the plaque. 'Tomás is the patron saint of Chichicastenango. I'm named for him.'

'And why did they choose Tomás as the patron saint?' I asked the boy.

He looked both ways to make sure no one was eavesdropping, and whispered, 'Because he can make soup out of bones.'

'OK,' I told Mickey. 'Let's go shopping.'

Out in the plaza, the funeral was gone, and the rain was easing. A few vendors were hopefully stretching wool blankets and ponchos on display racks. Tomás haggled with them like a grand master, and taught Mickey everything he ever wanted to know about bargaining Guatemalan style. 'Don't admire it too much,' he'd mutter of a blanket or a weaving. 'The man will get ideas. Tell him you don't want it. Tell him it's ugly.'

Or: 'He's asking ten. It's worth seven. Offer him half. Go to six, then walk away.'

Or he'd simply shake his head, whisper a figure, roll his eyes, hold up his fingers, and wink.

We ended up in a leather shop where an aged cobbler with a Fu Man Chu moustache made sandals, belts, and wallets by hand. The old man was exceedingly polite and very pleasant. He asked about North America and our families; Tomás even approved of his prices. Mickey brought *huaraches* and a tooled belt with a Quetzal painted on it, both of which had to be adjusted for size. While the cobbler worked, we waited. And while we were waiting, a local bus pulled up to the corner outside and started disgorging Indians. The top of the bus was bundled high with cabbages, onions, and yams, which the Indians were evidently bringing to market from the surrounding villages. The slick Ladino bus driver, glancing at his gold wristwatch and lighting a long filter cigarette, climbed up and started lowering down the cargo to his passengers to speed up the stop. He would hand one of the heavy bundles down to the outstretched hands of the Indian men and boys, several of whom then struggled to get the load on to someone's back. It went along like this for a few minutes. Then the bus driver, out of sheer malice, pitched down a crushing load of cabbages. It knocked one of the Indians flat on the sidewalk, skinning his face, ripping his pants, and tearing off his worn and flimsy sandals. As the Indians crowded around the fallen man to help, the bus driver jumped down, took the man's sandal when no one was paying attention, and tossed it underneath the bus. Then he climbed back on top of the bus and stood there laughing. The Indian men below slowly looked up at the Ladino, confused and frightened, as the injured man crawled to the wall of the cobbler's shop and sat there on the sidewalk, dazed and crying. It occurred to me in that instant that the Ladino's sadistic little prank might mean the Indian would have to go barefoot for a couple of years, maybe for the rest of his life. I'd lost track of how many men I'd seen in tears that day already, but I knew I'd seen enough: an uncontrollable urge came over me to send the Ladino's teeth down his throat.

'Don't get mixed up in this,' Tomás warned in his customary *sotto voce*.

'Why not?'

'*Los militares* – the military. If there's trouble, they'll come. They'll take the side of the bus driver.'

Mickey and I went out to the kerb, stared down the Ladino, and offloaded the rest of the produce ourselves. The Indians thanked us with claps on the back, then fled up the street with their vegetables. The bus moved on. When it was gone, we realized that Tomás, too, had taken off – afraid to get mixed up in any business that might bring the soldiers down on him.

We were as eager to get out of Chichicastenango as we had been to get out of Antigua, and Guatemala City before that. The downpour kicked again. We drove out of town, but only a few kilometres towards Santa Cruz del Quiché there was an army roadblock. The soldiers stood under the eaves of a house with Uzis in their hands and a machine-gun set up on the first-floor veranda. They checked out our papers then said the road to Santa Cruz was washed out: we couldn't pass. We had no way of knowing if they were lying, if up ahead the army was sweeping through Indian villages in the general's campaign to impose his will on Guatemala, shutting off the road back here so that no one could observe their counter-insurgency operations. Anyway, we could travel no farther into the highlands. The Quetzal's habitat in the cloud forests still lay north and west, beyond reach.

There was no arguing with the young guns. There was nothing to do but turn back, retrace our tracks to the capital, and try a more easterly route to the Verapazes.

'I look forward,' said Mickey, 'to another visit to the capital of this fine country.'

III

Into the Highlands

I

Back in the capital, my sleep was invaded by a thunderstorm that shook the hotel to its bowels; by aircraft taking off overhead, whose strained engines sounded as if they wouldn't clear the hotel roof; by a rumbling stomach, because we'd got back to the city too late to find an open restaurant; by a sense of high jinx brought on by our failure to get through Quiché; and finally, when I got a tenuous foothold in slumber, by a nightmare woven of all these hostile sounds and physical discomforts. We had gone back to the Residencial Reforma, a quaint guesthouse in a renovated nineteenth-century mansion, only a short way down the street from the American embassy. It had once been the private residence of General Justo Rufino Barrios, president of Guatemala in the 1870s and 1880s, who was known as *El Reformador*, The Reformer. I had seen an engraving of him on the face of the five-quetzal note, a melancholy old gent in a Vandyke beard whose long face and long nose and penetrating eyes seemed to share nothing with the perky Quetzal flying behind his ears on the note.

In my dream I was a passenger on one of those ageless Bluebird buses that are the standard public conveyance in Central America, picking up speed on a mountain road. It was night, and raining hard, and I was the only gringo on board. The other passengers were all ragged *campesinos*, wedged between their skinny wives and obese mothers and crying babies, three or four people for every seat. A few men were drunk and sick, others listening morosely to hopelessly mushy Mexican pop music on their tinny transistor radios. But for the most part the passengers were silently huddled together in the dark – like bad social-realist paintings of refugees. The bus was going faster and faster, skidding round corners at breakneck speed. And suddenly I realized we were out of control. The Bluebird bounced up and banged down, flew past shadowed *milpas*, met the boulders and the mud slides flying up at us, left behind the people waiting in the roadside ditches, all splattered and waving their fists. At last, only the black night and the sensation of crazed speed remained, the raindrops slanting uphill on the windows, distant fingers of lightning, and terrifying claps of thunder. Yet no one on board complained, or screamed, or tried to stop the driver. They were all passive and resigned to their fate, mumbling private prayers. By now I knew that this was how I was going to die, too, in a battered old Bluebird bus like the one I rode to school in as a kid, careening senselessly off the edge of a Central American night. Then I turned to look at the driver. He kept his face forward, but I could still make him out in the low garish light reflecting up from the dashboard. Dread struck me dumb, too frozen with fear to cry out. And the Bluebird bus dashed on into the driving rain, carrying another load of victims towards eternity. It was driven by a great, grizzled vulture wearing a busman's uniform . . .

I snapped on the radio and got the early news. The mayor of San Gabriel, Suchitepéquez Department, had been gunned down in the corridor of the town hall at three o'clock on the previous afternoon. The unknown killers had escaped, and the authorities had made no statement as to why anyone would want to shoot the mayor sixteen times with .357 Magnums. The mayor was fifty years old, and a Christian Democrat, the ninety-eighth Christian Democratic official murdered in the past two years. It was just what I needed to hear. My sense that the violence of the country had taken on a living will of its own was so strong that the tensions had begun to infect my psychic life like a virus. When the Mayans dreamed of the vulture, they said it meant that the rainy season was near. But if a vulture actually passed a Mayan house, it was an evil omen, a foretelling of death. Can a traveller's subconscious adopt the archetypes of the culture he passes through, I wondered.

I lay there in the wee hours with the rain crashing down, and the anxious sense of being watched. When I turned on the light and looked out of the jalousies, there was a light in the window across the tiny courtyard. A lone figure was standing there peering out, though whether a man or woman, whether looking at me or at the pouring rain, I couldn't tell. Perhaps a dream sharer. I lowered the blinds. As usual, Mickey was sleeping like a peaceful anarchist. Knowing there

would be no more sleep for me that night, I propped myself up in bed
and began scribbling in my notebook on the day's unrecorded events:

When we returned to the Natural History Museum this afternoon to check
in with Don Jorge Ibarra, the gate was closed. So was the door, but not
locked, so I let myself in. As usual, no visitors. Don Jorge sat inside his office
like a nut shrivelled inside its shell.

'*Pase adelante.*' He stooped up to shake hands. 'What can I do for you?'

He'd completely forgotten his promise to contact the coffee plantation
owner Schlehauf about permission for us to go Quetzal watching on his land.
Not only that: he'd forgotten me as well. His blank expression was the dead
giveaway. He seemed very withdrawn, as though contemplating empty space
inside himself. After a little verbal nudging on my part, it seemed to come
back to him. His secretary got Don Alfredo on the phone, and Don Jorge
made the arrangements. It turns out, Schlehauf is going to be up at his coffee
farm, Finca Remedios, in a little more than two weeks' time. We can
accompany him there. But that would mean returning yet again to Guatemala
City, something I'd like to avoid, so after Ibarra hung up, I suggested we meet
Schlehauf in the highlands instead. This was too much for the poor guy to
take: he sat at his glass-top desk in a kind of listless stupor, repeating over and
over again, 'What do we do? What do we do?'

I suggested he call Schlehauf back.

Don Jorge had his secretary dial again. This time he introduced her, though
she'd already been in the room with us for the past twenty minutes, and we'd
met on our previous visit. She was a pretty, petite brunette, deeply depressed,
with dark, red rings around her intense brown eyes that would have made
Lucia di Lammermoor in the Mad Scene look positively cheerful.

'This is my secretary, Margueritte,' Ibarra said. 'She's also my assistant.
She's very good, very intelligent, *mucho hábil* – much ability.'

The girl did not so much roll her eyes as lift them and let them drop again.

Mickey whispered, 'She looks like the *Mano Blanca* got hold of her last
night.'

In any case, instead of calling again, Ibarra agreed to my suggestion that it
would be better if Mickey and I drove over to Schlehauf's house to make our
arrangements face to face. After half an hour of rambling directions, scribbling
of addresses, sketching of maps, and so forth, though it was absolutely clear
the house was in a nearby zone, we got out of there. My farewell view of Don
Jorge: sunk into his chair in an exhausted slouch; spent.

I expected something grand from this plantation owner with the German
name, but Schlehauf lived in an unflashy development off the industrial section
of the Batres Road. The modern white stucco house had a low, locked iron

gate, no guards or dogs, and a mailbox marked *Zeitung*. His son let us in. The son looked about as German as José Ferrer. He and his dark-skinned girl-friend were listening to Donna Summer's records on the living-room stereo. Señora Schlehauf (or Frau Schlehauf – I was unsure how to address her) came out and gave us the twice-over, a thickset, unsmiling woman whose stony glance made us feel like encyclopaedia salesmen. On the walls were wondrous artifacts of the American tropics – a butterfly collection under glass, Mayan rubbings, wildlife photographs – but German was the lingua franca of the house. There was a young niece visiting from Germany who spoke a bit of English, and an older woman who may have been an aunt. They went back to their sewing while we waited on the living-room sofa.

After a while, Don Alfredo came out and shook hands. He was short, stocky, swarthy, wearing thick glasses, completely down-to-earth, refreshingly informal. He padded in in slippers and started talking casually in a throaty voice that sounded like a metal rasp.

'The Quetzal,' he began. 'Ah, it's the worst time of year to try to see the Quetzal. In this season the nesting is finished, and the birds are silent. Only by sheer luck you might see one. For example, the Quetzal nests in a tree hole of an old trunk, almost always in a small clearing. But because of the scarcity of dead trunks, sometimes a second pair will brood after the first pair leaves. In this case you might see them. Otherwise only by luck. I took some films of the Quetzals at Chelem-ha last year. Want to see them?'

In a few minutes we're ensconced around the dining table, watching Don Alfredo's home movies of the Quetzal on the minuscule screen. Even with the flickering projector and the bad light, the size and grandeur of the bird stand out magnificently. The bird waves its dazzling plumes, picks fruit from the trees, enters and leaves its nest hole. What amazes me is not the plumage, however, which is poorly shot in any case, but rather how elegant the Quetzal's movements are, as if each one had been choreographed. In one sequence the bird is perched on a branch, facing away from the viewer. Repeatedly, it does no more than flick its tail, and the wave ripples down the entire length of the plumes. Like a living jewel!

After seeing the films I told Schlehauf that we wanted to go to Chelem-ha more than ever, despite the odds against seeing the bird. I think he recognized the fanatic can-do spirit by which Americans seek everything, from birds to God, and realized that against such naive positivism rational argument was useless. He readily consented, and brought out a hand-drawn map decorated with kitschy alpine cottages, stick figures of peasants under banana groves, and the roads stitched out of red yarn. The title was calligraphed: 'Chelem-ha, Refuge of the Quetzal'. We went over the route to the *finca*, sitting on the living-room sofa drinking black coffee with sugar. It looked as though the mountain itself stands about twenty kilometres north of the plantation. The mountains of the vicinity, Don Alfredo said, are about 2,000–2,500 metres high, Chelem-ha itself, at 2,200 metres, all cloud forest. It seems the main part of the refuge is a curving ridge at the northern limit, falling off on the north side through jungle down to the road that leads to Cobán. A farmed valley runs up the bowl towards the ridge. Schlehauf also brought out a photo scrapbook, showing us pictures of how, year by year, more of the forest was being burned and planted in *milpa* by Indian squatters.

After a bit of negotiation, we set a date for meeting him at Finca Remedios, went over the complicated route again, and took our leave.

I liked Alfredo Schlehauf. With his guileless manner and apparent interest in plants, birds, and butterflies, he reminded me of the Swiss and Germans I'd met in Costa Rica, who formed an educated, liberal middle class and made for social stability. He seemed to have inherited the best qualities from a *mestizo* background – the scientific interest in nature from his German side without the German formality, the earthiness of the Central American *campesino* without the arrogance of the Ladino elite.

I slept only briefly before dawn broke the colour of a dirty dish-rag off the hotel veranda. A Blue-grey Tanager flitted through the shrubbery, hunting for its breakfast of berries. Even though among the more muted of its brightly coloured family, the little Blue-grey was easy to follow with the naked eye. For several minutes I watched it work the hotel's landscaping, where it seemed to be making out better than most humans in the city streets. Then a hummingbird appeared, zinging too fast to identify through the papery clumps of bougainvillea flowers. Even such tame and tenuous observations gave me new heart that ornithology in Guatemala might still involve something better than roadblocks, rain, and Black Vultures. By 8 a.m. our little golden Toyota four-wheel was sweeping confidently through the grotesque suburbs on the opposite side of Guatemala City. We had decided to cruise up to the highlands on a north-east slant along the old Cobán Road.

At the first military checkpoint, a tollbooth just outside the city, it didn't go too badly; at least the soldiers didn't point their rifles. They pointed their fingers instead, to the spot in the road where they wanted us to line up behind a dozen other vehicles. As soon as we had done so, they promptly forgot about us, and went back to schmoozing in the shade. From every direction women swarmed around the shiny new Toyota, bearing on their heads baskets of sliced watermelon, papaya, mango, and roasted ears of corn. The kids were waving bags of peanuts, cookies, chocolate, caramels, and Chiclets. One wanted to sell us a lottery ticket, another had the morning papers. We breakfasted on sticky mango in a plastic envelope and two bags of raw peanuts. When everyone had had a go at us, they retreated to watch us from a safe distance. We simply sat waiting for half an hour, then an hour. Nothing happened, as if by custom. No one else was getting through, except for a black limo with government plates that didn't stop at all. The soldiers stood under the shady eaves of the guard booth, ignoring the halted traffic. So we sat in the melting asphalt of the road like everyone else, remembering with a sudden and unexpected fondness the exact-change lanes of the turnpike tollbooths in the States.

Eventually, one of the ragged kids offered to run our documents into the guard booth. He raced back a minute later and told us they wanted to see us inside. I gave the kid a tip of one quetzal. Inside, behind a screened booth, sat a soldier-clerk in spectacles, like a priest in a confessional. The soldier never raised his eyes. He drew a long, pink slip from the stack on the desk, wrote down our passport numbers and auto registration number in his book, asked where we were headed, and our purpose ('Cobán.' 'Tourism.'), charged forty centavos, stamped the pink slip 'Paid', handed it to me along with our documents, and bade us enjoy our stay in Guatemala. We were on our way in about thirty seconds.

I was in the passenger seat, working myself into a fine verbal tantrum about how nothing, not even the simplest civil function, could get accomplished in Guatemala without barefooted kids, when the Toyota suddenly veered, as if drawn by a magnet, and Mickey pulled into the most excruciatingly well-run Exxon station you could ever hope to encounter. A young man in a spic-and-span service uniform greeted us at the junction of the station lot and the road, ushering us towards the pumps with a courtly gesture. Then two youthful colleagues attached themselves to either side of the car, to guide us gently forward. Five others snapped into a flurry of activity, beckoning to us from the precise spot where they wished us to pull up under the metal awning, with energetic arm waving and whistling. As soon as we had stopped, they immediately fell to with buckets and sponges, scrubbing the Toyota, which was only superficially dusty. Meanwhile, the station owner, in a clean white shirt and dark slacks, turned up the marimba music he played over the PA for his customers – perhaps I should call them clients – and then catapulted out of his office to take personal charge of this sacred act of car washing – a freebie, as he explained, in honour of the station's 'Grand opening'. He was only sorry we hadn't arrived in time to make use of the automatic car-washing machine. But it only functioned at 'specified hours', he affirmed, pointing down the lot to where a little girl of about eight years sat before the machine on an overturned plastic bucket, holding the pull cord that operated the contraption, though only at the 'specified hours'. Nonetheless, other than that, we got the full treatment. One of the owner's minions was under the hood checking the oil, while a second gauged the water, and a third descended with an air hose for the tyres: he looked confused when I informed him we'd only recently got the car from the agency, and the tyre pressure was OK. The station scribe also stood by, pad in hand, awaiting our gas order.

I told him to fill it up with regular.

He carefully noted my words, and transmitted the same orally to the *muchacho* standing at the ready, pump nozzle in hand, who waited for yet another employee to locate and remove the gas cap. The entire operation consumed the labour of fourteen able-bodied men, not counting, of course, the station owner himself, who was really only overseeing the fill-up, nor the little girl, loyal to her post beside the automatic car-wash machine. Was this some sort of Central American parody of gringo car culture?

Ten minutes at the Exxon station was enough to restore our humour. It was like a stop at a health club. We shook hands with everyone, and promised the owner to return for our free car wash. I didn't know the Spanish word for 'rain check'. Then we hit the road again.

2

Up and down, round and round, the countryside never stayed the same more than a few miles at a stretch over Guatemala's diverse geography. Granite heights that looked clawed by blind and angry titans pitched into patches of lowland rain forest, blooming in erotic dishevelment: a restless topography that would be all things at once, display all seasons, demonstrate all climates, contain the living evidence of all ages, conjure all the varied forces of nature. It was like an entire continent stuffed as in an expertly packed suitcase into a country the size of Massachusetts. At length we descended from mud to dust, into a dry scrub plain. At El Rancho, where a mammoth electric-power-station complex under construction blotted the sky with sandy yellow grit, we left the Ruta Atlántica, which leads northeast on to Lake Izabal and eventually to Puerto Barrios on the Caribbean coast, where the banana boats load. North on the left-hand fork we began to climb once more, this time through desert. This one crept south-east on our right side into the Guatemalan Oriente, a region infamous for its withering climate, venomous snakes – as well as for its short-tempered, violent Ladinos who are said to leave their hammocks only to cut each other up in machete fights.

The wayside flowers disappeared, along with all signs of cultivation. Stiff, forbidding clumps of prickly pear and tall cereus cacti stood guard, nasty plants bristling with hostile spines, devoted so ruthlessly to their lonely battle for survival in the pervasive dryness that their very morphology seemed to have shaped itself into an utter disdain for interaction with other living things, a harsh contempt for natural cooperation. Who needs shade, breeze, or rain they seemed to say. What use have we for mates, companions? Doggedly prying apart a crack in the rocks to jab a root down, they behaved like steely loners, suspicious and intolerant of intruders. It was cactus country from here south to the Salvadoran border.

Yet the arid region was brimming with contrasting bird life, as bright and bouncy as the cacti were aloof. Troops of screaming green parrots swooped across the road, flying in precise, tight formations towards the verdant edges of a tributary of the Río Motagua, a swollen, muddy stream we could make out a mile or so below the road. Streak-backed Orioles stood out like licks of flame against the muted grey-greens, and low in the scrub tiny Blue-black Grassquits cheeped and jumped. At the ninety-two-kilometre marker, I cried, 'Stop the car!' Beside the road a pair of lovely motmots perched in a small scrub oak atop a rocky rise. I'd only glimpsed a flash of their almost turquoise plumage, and their fantastic 'racket-tipped' tails, a bizarre accoutrement formed when the weak middle barbs of the motmot's plumes break off through wear or preening, leaving bare spines eight inches long, tipped by short, rounded, stronger barbs that give the bird's tail feathers the look of a tennis racket. The motmots perched serenely, giving me time to thumb through Hugh Land's *Birds of Guatemala*. With field-glasses I could pick out an iridescent blue-green ring around the brow, the black throat spot, and a cinnamon shading of the crissum, leading to the conclusion that these were Turquoise-browed Motmots, 'common to the arid lowlands'.

It was past noon when we pulled into the next checkpoint. While the soldiers studied our documents, two men in yellow oilcloth slickers and vinyl helmets, looking much like your average aliens, trudged up from nowhere to announce they were from a disease control unit, and intended to fumigate the car with DDT. They carried portable spray machines strapped to their backs. As we couldn't very well continue on our way, at least until the fog dissipated, we decided to stop for lunch. It was a truck stop, anyway. The roadside opposite the guardhouse was lined with the squalid little Central American eating places known as *comedores*, one right next to another. They were not by any stretch of the imagination proper restaurants, which are almost unknown in the Central American countryside, only simple, open-fronted, dirt-floor shacks, with a counter and bench out front, or a makeshift table, though most had only a stand-up counter. Sitting down to eat is a luxury to most Central Americans.

We went up to the closest stand. Inside the hot kitchen three women – three generations of the same family by the looks of them – toiled over their smoky charcoal brazier. The grandmother, kneeling on her haunches, was scooping patties of unleavened corn dough from a basketball-sized mound and patting them expertly between her palms into *tortillas*; they were too poor to own a handstone and quern, let alone the comparatively high-tech manual *tortilla* presses so common in Mexico. Behind her, the middle-aged woman baked the flat cakes on a griddle, and also attended pots of sinister-smelling stew and beans; her daughter worked as the waitress. They served up the typical and unvarying *comida* of *tortillas*, black beans, and green chillies, to be washed down with a Coke or a Fanta. Meanwhile, outside the shops, younger daughters tended their even younger naked siblings; mangy dogs curled up in the dust; and a huge sow ambled up and down the line, rooting around for slops.

Mickey took one long, gloomy look at the place: I could see in his expression that a little health commissioner was running around his head blowing a whistle. He ordered a Fanta, and said, 'How can you eat that stuff?'

'You've got to eat something,' I said lamely. Then, 'There won't be anything better farther on.'

What with the swill running in the narrow canal underfoot and a cloud of DDT enveloping us each time the boys in yellow gave a toot on their fogging machines, it didn't look especially appetizing. Yet there really was no choice, and there would be none down the road. It was either eat or fast, or try to subsist on Coke and Fanta. I could only take comfort in the fact that the food at most *comedores* is, at least, usually fresh, as well as cheap. From time to time you encounter the odd surprise that makes eating in Central America something more than a bare necessity: for example, the woman who makes wonderful *chiles rellenos* and stuffs them into a roll sotted with ketchup, mayonnaise, and hot sauce, or a bowl of clotted sweet cream to liven up the beans.

But here, as at most *comedores*, you stuck with the plain, maize *tortillas*, which Central Americans eat twice or three times a day seven days a week, year in and year out. Standard fare, fast food, ubiquitous comforter of the body and soul in a land where everyone is on the move all the time, either eating in *comedores* or carrying their food with them. How emotionally stabilizing it must be for the Guatemalan to know that wherever he goes, he will find the customary nourishment. The same shape, colour, taste, smell, and even more. For the Indian tradition says that eating *tortillas* makes you human – literally makes a man out of you – or so the legend goes. The Popol Vuh, for instance, while lacking the theme of original sin, puts maize at the very centre of human creation. The first men on earth behaved badly, to the expected anger of the gods, but not because of forbidden carnal knowledge. The gods themselves had made the mistake of fashioning men out of mud:

Of earth, of mud, they made man's flesh. But they saw it was not good. It melted away, it was soft, did not move, had no strength, it fell down, it was limp, it could not move its head, its face fell to one side, its sight was blurred, it could not look behind. At first it spoke, but it had no mind. Quickly it soaked in the water and could not stand.

And the Creator and the Maker said: 'Let us try again because our creatures will not be able to walk nor multiply. Let us consider this,' they said.

It was a matter of the right stuff: on a subsequent try, Gucumatz, the creator, fashioned men of *tzite*, of maize: 'Thou corn; thou, *tzite*, thou, fate; thou, creature, get together, take each other.' A prayer which reminds us that all life begins again with pollination.

The eating of corn *tortillas* had varied little in thousands of years of Mayan history, back to the legendary era when the king and lawgiver Quetzalcoatl first domesticated the wild highland grass called by the Mayas *zea Mays*, and gave it to his people as the most enduring gift, civilization. For just as Western civilization in the cradle of the Near East began with the domestication of wheat, goats, and sheep, American civilization may be said to have curled its way up the corn stalk. At first, most likely, in the transition out of nomadism into settled village life based on the cultivation of a staple foodstuff. Later, with the advance of farming techniques based on the dibble, the hoe, and the machete – all still utilized today – came 150 varieties of cultivated maize, came surplus, trade, and an indigenous culture affluent enough to support not only a priesthood and nobility, but a class of sophisticated, professional artisans who took their inspiration from the cycle of maize agriculture. Later still, it may well have been exhaustion of the soil from the monoculture of corn that precipitated the sudden, still unexplained collapse of classical Mayan civilization. The tropical and subtropical portions of the Western hemisphere had proved neither well-enough endowed with game to support an expanding population, nor sufficiently provided with open plains suitable for the grazing of the larger domesticated animals. Yet according to numerous studies, it's the modern introduction of beef cattle that's chewing up so much of Central American lands previously under maize cultivation. Dr Moisés Behar, head of the UN's Institute for Nutrition of Central America, has made the statement that the pre-Columbian Mayas ate better than the Central Americans of today.

I chewed my *tortillas* slowly. Like water itself, the *tortilla* is all tastes combined, and no taste at all. It is history, and outside of history. All of Central America – the great pyramids, the unread stone lintels, the burning skies of volcanoes and the rumbling of earthquakes, the old gods, the baking sun and shattering rains, the centuries of conquests, and exploitation, rebellion, but, most of all, the tranquil, stolid, reticent personality of the Indian peasant – seems to be released from these chewy, bland, slightly smoky and rancid cakes. You taste in them the Indian peasant's tenacious loyalty to his roots, his isolated and ethereal temperament as he scratches the thin, poor laterite soil of his *milpa* – a dogged farmer growing this hardiest of crops, on increasingly unyielding earth. Over every morning fire they are toasted for breakfast. They serve as edible plates for the evening's supper. In a folded kerchief the *campesino* carries them up into his *milpa* for his noon-time nourishment, and the women tuck their small stacks of *tortillas* into their bosoms as they walk the roads to market. To call the *tortilla* the staff of life does not do it justice. To eat the *tortilla* is to accept the wafer of sacrament for the isthmus of middle America.

Past fumigants and soldiers, the road straightened out. At our backs, the oppressive heat skirmished with the moisture-laden clouds of the mountains. We rose rapidly, and once more the prospect changed completely. We had entered the department of Baja Verapaz, headed north-north-east for the high elevations, where the cloud forests form the traditional range of the Quetzal. Here at the middling elevations, there were few vast panoramas, and no dizzying geologic displays. The afternoon sun dispersed through a region of bucolic miniatures. The light seemed to waft upwards from the earth like steam, making the thick foliage tingle and glisten. We began to pass small farms – *granjas*, they are called – hidden in groves of floppy-leafed banana trees, surrounded by thick green corn stalks. Only now and again could you catch a glimpse through the enveloping vegetation of a sway-backed red tile roof, a wattled hut roofed with thatch, or a shady porch, where an Indian mother rocked her kitten in a tiny hammock. The honeyed smell of warm orchids, oleander, hibiscus, and allamanda rose up through the door-yard *milpas*, which were double planted beneath the maize with beans and squash vines, forming natural garden walls. The *granjas* sequestered an interior life, a tropical version, it occurred to me, of those seventeenth-century Netherlandish landscapes, masterpieces of suffused light and intimate country colours. And the *granjas* reminded me, too, of a phrase read in *The Annals of the Cakchiquels*, the post-Columbian Mayan record of the Spanish conquest, written collectively by sixteenth-century Indians who had learned Spanish in Sololá, a Guatemalan town near Lake Atitlán. Whenever the narrators of *The Annals* wished to describe how the Indians fled to refuge in the mountains from the rampaging conquistadors, they repeated the redolent line 'under the trees, under the vines'.

> We scattered ourselves under the trees,
> under the vines, oh, my sons!

The *granjas* seemed to sigh sleepy peace. We spun by in a drowsy daydream, under the illusion that there is no more tranquil spot on earth than here in Central America, under the trees, under the vines. Perhaps it's true, as Marxists say, that these small holdings are only enough land for the *campesino* to starve on when there's no wage labour to be found on the large plantations. Yet it's hard not to surrender to this sense of utter privacy, in a landscape of complete seclusion, turning out towards the traveller the face of Central America one can grow so easily to love: a rural poverty ameliorated by the colour green, the chaste and simple country life among plants and animals. Perhaps the traveller's belief that only those who stay put, close to their birthplace, nourishing and taking nourishment from the land, can keep the sun up in the sky. And the frailest hope that the sewer of history might flow by without fouling these secret pockets of true peace, which is the meaning of the name *Verapaz*.

3

Therefore upon learning from them of their ill will towards His Majesty's service, I burned them for the sake of the peace and welfare of this land, and I gave orders that the city was to be burned and razed to its foundations.

Pedro de Alvarado
in a dispatch to Hernando Cortés

Then Tunatiuh asked the kings for money. He wished them to give him piles of metal, their vessels and crowns. And as they did not bring them to him immediately, Tunatiuh became angry with the kings and said to them: 'Why have you not brought me the metal? If you do not bring with you all of the money of the tribes, I will burn you and I will hang you,' he said to the lords.

The Annals of the Cakchiquels

Oh, how many orphans did he make, how many families did he rob of their sons, how many husbands did he deprive of their wives, how many women did he leave without husbands, how many married women did he adulterate, how many virgins did he ravish, how many did he enslave, how much anguish, how many calamities did the Indians suffer because of him! How many tears were shed, how many groans were uttered, how many sighs did he provoke, how many were condemned to eternal damnation, and not only Indians in great number, but also unfortunate Spaniards whom he encouraged in wickedness, and who assisted him in the committing of so many heinous and abominable murders. I do beseech God to be merciful on their souls and be satisfied with the vile ending He gave that tyrant [Alvarado] . . .

Fray Bartolomé de Las Casas,
The Tears of the Indians, 1542

The Guatemalan Highlands have not always been known as an area of true peace. On the contrary, in the first decades of Spanish conquest the region was called *Tierra de Guerra* – land of war. On three separate occasions the Spaniards returned from attempts to subjugate the mountain Mayans 'holding their heads' in woe. Pedro Alvarado never himself travelled in the highlands, lured south as he was to search for Cibola, the seven cities of gold. Yet the story of Verapaz must be treated as part of the larger story of Alvarado's conquest of Guatemala, not only because the area's subsequent pacification by Bartolomé de Las Casas was inspired in part by the bloody devastation of Alvarado that preceded it, but also because many of the Indian refugees from Alvarado's campaign fled to these same mountain strongholds. This accounts in part for the fact that even today three of the major Mayan dialects – Quiché, Kekchí, and Cakchiquel – are still spoken in the disparate villages of the Verapaz highlands. In the turbulent epoch of the conquistador begins the long conflict between displaced indigenous and Spanish cultures that has gone on in Central America for nearly five hundred years, and still goes on today.

It has been suggested that Guatemala's founder was a better man than historians have made him out to be, and that had Pedro de Alvarado only lived to write his own *probanza* – the self-serving memoirs of service to the Crown that ageing conquistadors were wont to commit to parchment – he might have countered, or at least softened, the wrath-of-God image Las Casas pinned to his armour in a flood of theological diatribes. It's true, of course, that in many respects Alvarado was no different from the other conquistadors. He shared precisely the same background as Vasco Núñez de Balboa, Francisco Pizarro, and Hernando Cortés: an aristocratic family, from Estremadura in the Spanish Pyrenees, without, as they say, a pot to piss in. Alvarado's people had lived on the front line of Christianity's defence against the invading forces of the Muslims for nearly eight centuries. To them, the Crusades were no far-flung oriental adventure to free a distant Holy Land, but rather a war to liberate their own Spanish homeland, where the Berbers and the Moors long dominated. So many generations of religious/military conflict with the Arabs could not help but shape, and warp, the Estremaduran temperament. What grew in the bitter soil of occupation was combativeness, stubbornness, and the fierce martial pride of a ruined nobility, vanquished by the foreign infidels.

In much the same measure that the Muslims and Jews who settled in the Iberian peninsula devoted themselves to agriculture, science, the arts, and commerce, the indigenous Catholic nobility spurned such pursuits as works of the devil. Their singular devotion was to the warrior's trade. A particularly stinging contempt, saved for anyone who worked the land, would carry over, little changed, to the New World with devastating effect. For the conquistadors never seriously considered the prospect of staying, colonizing their discoveries, building their conquests into a new agrarian society. Cortés spoke for all conquistadors in his outright refusal of a royal land grant in the New World: 'But I came to get gold, not to till the soil like a peasant.'

Yet just as they were finally succeeding in turning back the Muslim wave, the Iberians saw the rest of Europe progressing rapidly into the capitalist era, on the strength of those same agricultural and commercial skills which they themselves held in disdain. It must have seemed to them an ultimate injustice. Why should the other Europeans advance, when it was the Spaniards who had saved Christian Europe? Centuries of sacrifice and heroism on the battlefront had convinced the Spaniards that they were God's personal garrison, His chosen instrument, destined not merely to win back their own territory, but to gain the position of eminence among Europeans a just God would surely reward them with. While warfare and salvation, then, were integral to the conquistador's inheritance of faith, it was the lust for gold that came to connect his militant faith to his worldly ambitions: only the rapid accumulation of wealth might elevate the Estremaduran to his fitting station. It would be as wrong to doubt the sincerity of the sixteenth-century Spanish Catholic's belief in avarice in the service of the Lord as it would be to doubt the nineteenth-century Protestant's spiritual calling to work and save. By no means did the secular reward of wealth contradict the spiritual struggle. If these pitiless militaristic fanatics, these obsessively romantic mountain aristocrats, were prepared to risk their lives and abandon their families for the gold and glory of exploits into the unknown, their consciences were absolutely clear in the belief that they were on a sacred mission. Emperor Charles V vowed to gain the pope's personal indulgence and absolution for their mission before the conquistadors ever set sail for the Americas. Thus they believed themselves assured by the highest spiritual authorities on earth that their wars of conquest would be just wars.

The feudal morality, iron discipline, and fanatical monotheism with which the conquistadors moved across the New World could hardly have been less appropriate to the tropical, pantheistic landscape: in retrospect, the conquistadors' indifference to the natural situation of the Americas remains perhaps their greatest intellectual feat, admittedly a negative one. They wore the same heavy armour marching through the feverish humidity of the swamps as through the chilly mountain nights. They often failed to differentiate between one indigenous tribe and the next. They rarely stopped to wonder at the monuments of an American civilization in certain respects more advanced than their own, and had no concern whatsoever that they were destroying one of humanity's great written literatures in the native chronicles – so much satanic trash. They took scant notice of the miraculous variety of plant, bird, and insect life (keeping a sharp eye only for the exotic, which could be paraded through Spain as a selling point for future expeditions). In addition, they remained stubbornly unaware that the Indians did not use gold for money – most Meso-American tribes kept track of their trade in cocoa beans – and remained deaf when the Indians said there was little gold. Though most of the time, the Indians weren't lying. For all it mattered to the conquistadors, they could as well have been campaigning in China. Some, of course, thought they were.

On the other hand, even in a historical moment of mass delusion nourished on blind faith, Pedro de Alvarado rises above the commonplace to display several of the special qualities of the psychopath at work. Alvarado first comes to notice as the handsome, foppish, cunning young captain who almost singlehandedly wrecked Cortés's expedition to Mexico. With his noble bearing, athletic prowess, natty dress, icy blue eyes, red hair, and pointed red beard below curling red moustaches, Alvarado inspired wonder among the Aztecs, who soon dubbed him 'Tunatiuh', the Sun. His own comrades also noticed Alvarado's solar traits: one moment he would be talking and laughing hysterically, the next throwing a violent temper tantrum.

The intrigues that touched off war between the Spaniards and the Mexicans will never be known for certain. What we do know without doubt is that when Hernando Cortés left the Aztec capital of Tenochtitlán in Alvarado's command, to go and deal with the threat of Pánfilo de Narváez's arrival on the Mexican coast, the city was in a state of relative calm, with the Aztec Emperor Moctezuma under house arrest. A few days later, all hell had broken loose. According to the Aztec account, as soon as Cortés had left, Alvarado gave permission to the Aztec priests to put on the feast of Huitzilopochtli, in which the sacrifice of human hearts played a prominent part. But when the Aztecs were dancing before their pyramid temple, Alvarado suddenly signalled his men to attack. The fully armed Spanish soldiers sprang into the square and began to butcher the unarmed celebrants. Afterwards, they stripped the bodies of jewellery and mutilated the corpses. Within hours, Tenochtitlán was in active revolt. Cortés was forced to hustle back, and confronted Alvarado in a state of white fury. Alvarado defended his order to ambush the Aztecs with his usual verbal adroitness. He had intercepted certain secret communications, he said, indicating that the natives were planning to attack the Spaniards. So he decided to launch a pre-emptive attack, allowing the natives to perform their religious ceremony as a ploy to catch them off guard.

It wouldn't wash. According to William H. Prescott's *Conquest of Mexico*, Cortés denounced Alvarado in the most strenuous terms. 'It was all wrong and a great mistake,' he said. 'You have done badly. You have been false to your trust. Your conduct has been that of a madman.' Then he turned his back on the junior officer and walked away – a grave display of disrespect. Had Cortés not been in dire need of his best fighting captains to face the Aztec insurrection, Pedro Alvarado's career as a conquistador might well have ended in irons – or perhaps on the gallows.

Nor can we know why, after the Spanish victory in Mexico, Cortés charged Alvarado and his brothers to undertake an expedition south to Guatemala. Maybe he needed a vacation from the Alvarado boys, whom one historian has described as 'terrorist gangsters, accomplished plunderers, slave-hunters, and extortioners'. Maybe Cortés thought a long trip through fever- and snake-infested jungles would do the Alvarados some good. In any case, on 6 December 1523, Pedro Alvarado found himself marching south towards Central America, at the head of a column of 120 Spanish horsemen, 300 foot soldiers, and several thousand Indian slaves and forced-labour warriors. Cortés explicitly enjoined him to win land for Spain by 'peaceful means', and to treat the people 'with loving kindness'. Anyone as well acquainted with Alvarado's record in Mexico as Cortés was should have had little confidence that the captain's southward march would be anything short of catastrophic for the native inhabitants.

In the event, Pedro Alvarado's expedition to Guatemala stands as one of history's staggering butcheries, an appalling genocide which Las Casas estimated, in his hyperbolic tract *A Very Brief Account of the Destruction of the Indies*, as costing the lives of five million Indians. Surely an exaggeration, but it does give an idea of the monstrous scale of the slaughter. Operating far from any power to restrain him, 'Tunatiuh Avilantaro', as the Guatemalan tribes mashed his name, commenced two years of ceaseless bloodshed. Unlike Mexico, where the power of the Aztec empire was concentrated in the hands of a single demigod, the Mayan tribes were politically and religiously decentralized, and geographically scattered. They were the remnant peoples of the sophisticated Mayan civilization that had collapsed more than five hundred years previously. Although 'decadent' and 'degenerate' would not be accurate terms to describe their culture, there is little doubt that the Mayans of the conquest era had devolved far from the heights of Classic Mayan civilization. For the most part, they were marginal country folk, who had slid into tradition-bound agrarianism, moving neither forwards nor backwards, often fighting among themselves.

Without following each vine of the dense jungle of his campaign, Alvarado's general strategy followed that of Cortés in Mexico – divide and conquer. Upon arrival in a district, he would make a temporary alliance with one tribe against its traditional rival. Together they would vanquish their common foe. No sooner was this accomplished, however, than Alvarado would turn on his allies with demands for gold, tribute, and slaves that could not be met; whereupon a new round of slaughter would begin. Thus the Cakchiquels, who had sent emissaries of peace to Cortés in Mexico, were enlisted to fight the Quichés, and the Zutuhils to fight the Cakchiquels. Burning cities as well as crops, to deny the population everything, Alvarado alternately starved the Indian warriors under his command, and then, so Las Casas charged, 'permitted them to eat the flesh of the Indians that they had taken in war ... suffering children to be killed and broiled in his presence'. The Indian chiefs were extorted for gold, taken prisoner, burnt alive. The women were ravished, impaled, or consigned to the Spanish soldiery in lieu of wages. The children were bashed and baptized, in no certain order, and the survivors enslaved, branded, and set to work mining gold for the conquistador. Where Cortés had refrained from using the infamous 'dogs of war' to hunt and kill natives, Alvarado set them loose. Thus in the very act of founding the Guatemalan nation, Pedro de Alvarado introduced the practice of war against the civilian population, initiating the tradition of genocide that has bedevilled Guatemala as its greatest shame down to the present day.

In this way, in the course of two years, Pedro de Alvarado became indisputable tyrant of the lands of the Quetzal, including present-day Guatemala, El Salvador, southern Mexico, and part of Honduras. But as the gold of the viceroyalty of Guatemala proved scarce, he remained restless. While the allegations of Las Casas – and of Cortés himself, who officially rebuked Alvarado's conduct in his reports – made their way slowly through the Spanish legal-ecclesiastical system, the conquistador did not stand pat. Alvarado left the mopping-up operations in Guatemala to his brother Jorge, and set off on new exploits. He visited Spain, successfully defending himself against the charges by an expedient reassertion of loyalty to the Crown; returned to raise a slave army of Guatemalan Indians; and attempted to beat Pizarro to the imagined riches on the northern margin of the latter's Peruvian empire. Pizarro's captain Belalcázar reached Quito first, but by bluff and bullying (he had ships, soldiers, and knew how to make trouble) Alvarado was able to extort from Pizarro a thousand pounds of pure gold just for leaving Peru. He returned briefly to Guatemala, sailed again to Spain, came back with his new bride, Doña Beatriz, but only long enough to install her at his seat in Santiago de los Caballeros near the present site of Antigua in Guatemala.

And each time Alvarado landed in Guatemala, fresh sorrows came to the Indians. In one infamous incident, Alvarado apparently lost his temper at the Cakchiquel *Ahtzib Caok*, or lord, who had come to greet him on his return from Spain. He beat the old chieftain so badly that the man died of the injuries within twenty-four hours. Later, Alvarado hanged the last tribal chieftains of the Quiché and Cakchiquel tribes. When Pedro de Alvarado died in 1541, at the age of fifty-six, victim of a horse crushing him while he was putting down an Indian revolt in Guadalajara, he was still hard at it, raising yet another army to go for the gold one more time, in the Spice Islands. 'Tell us where it hurts you,' asked his faithful followers. 'In my soul,' said Pedro de Alvarado. In the immediate wake of his death came the bizarre episode of his twenty-one-year-old widow Doña Beatriz's reign, which ended after twenty-four hours in the total destruction of the Spanish colony at Santiago de los Caballeros by a combined earthquake, volcanic eruption, landslide, and flood – a veritable rampage of the old Meso-American gods celebrating Alvarado's demise. Curiously, the native annals omit news of Alvarado's death entirely, and report the subsequent events with a restraint that is difficult for those with a Western sense of revanche to fathom:

On the day 2 Tihax there was a landslide on the volcano Hunahpu; water gushed from inside the volcano, the Spaniards died, and the wife of Tunatiuh perished.

4

Our grandfathers and fathers died together. And great was the stench of the dead . . . the vultures devoured their bodies.

The Annals of the Cakchiquels

While vultures roamed profitably over the newly conquered territories, Bartolomé de Las Casas's protests against Alvarado's savage methods were not met with indifference in Spain. The priest, later to become known as the Apostle of the Indies, worked tirelessly on behalf of the natives, trying to convince both Church and Crown that their own interests were poorly served by the likes of the conquistadors. The natives of the New World were human beings, he argued, and as such were capable of accepting the True Faith, if only it were advanced through purely peaceful means. Las Casas was as much inspired by this sweet vision of universal religious teaching as the conquistadors were by the chimera of El Dorado. No doubt, he was just as fervently committed to his pacific mission as they to their just war. Las Casas wielded his pen as zealously as Alvarado his sword – and in the end, both believed firmly in the ultimate power of his chosen weapon. Significantly, the Apostle of the Indies had in-fluence at the Vatican, as well as sympathetic ears at the court of Charles V.

By 1533, Bartolomé de Las Casas was leaving behind ten years of self-imposed monastic life on the Caribbean island of Hispaniola, where he'd retreated after the bitter failure of his attempt to establish a Christian colony called Tierra Firme on the coast of Venezuela. The débâcle in Tierra Firme, where Franciscan friars had successfully preached Christianity to the natives only to have their good works wiped out overnight when renegade Spaniards moved in on slaving raids, had taught Las Casas an important lesson, namely, that the crucial precondition for any patient, tranquil appeal to the natives' religious reason was keeping the conquistadors away from the natives. With psychological insight far in advance of his time, Las Casas argued in sermon and text that the Indians could not become persuaded to the faith

as they wander about, hidden and scattered through the woods and forest . . . stunned and frightened by the unbelievable terror with which their oppressors have filled them . . . so shattered with fear that they want to hurl themselves headlong into the deepest caverns of the earth to escape the clutches of these plunderers.

From Mexico, Las Casas submitted to the Spanish Council of the Indies a new, detailed plan for the peaceful conversion of the Indians, asserting that 'every day you will see the fruits of this just government. This, gentlemen, is *just* government. This is the right path.' In order to ensure the complete separation of Spaniard and native, he chose a site for his experiment so remote, hostile, and seemingly impenetrable that Spaniards would have no desire to intrude. And so the year 1536 found the Apostle of the Indies in Guatemala, urging Alvarado's stand-in Maldonado to allow him to demonstrate his method in the infamous highlands of the Tierra de Guerra. From the pulpit in Santiago, Las Casas read passages of his work in progress, titled, 'The Only Method of Attracting All People to the True Faith'. His call for a ban on violence, for conversion by example, for return of lands stolen from the natives in unjust wars, aroused the astonishment, to put it mildly, of the hard-bitten veterans of Alvarado's campaign, who made up the congregation. Go right ahead, *padre*, they urged him contemptuously, go up to Rabinal and teach the *caciques* the 'peace and love' of Christ. They were sure the friar was a raving madman. He would either be served up straight away as highland stew, or taught a lesson he would not be likely to forget. For this reason, no one offered serious opposition to his project. Las Casas made two formal requests of the governing Spanish authorities. First, that the Indians he converted should not be divided among the Spanish as slaves in the usual fashion, but instead be put under the direct supervision of the Crown. And second, that all secular Spaniards be legally restrained from entering Tuzutlán province for five years. Maldonado, more the administrator than Alvarado in any case, yielded on both points.

With this pact concluded, Las Casas and three Dominican associates set to work composing a set of hymns, more like ballads, in the native dialect, recounting the creation, the exile from Paradise, and the life of Jesus – vernacular gospels in musical form. They next enlisted four Christian-Indian merchants, accustomed to trading in the highlands, and familiar to the people there. Verse by verse, the traders were patiently taught the friars' songs, until they knew each one by heart, and could sing them, as Las Casas noted, 'in a pleasing manner'. In August 1537, the merchants set out on the road, while Las Casas and the Dominicans remained in Santiago, praying, fasting, and, no doubt, worrying. The merchants, as instructed, went directly to the great *cacique* of Tuzutlán. After they were done trading, they called for the stringed *teplanastle* instrument, and proceeded to sing: the novelty of the situation requires no elaboration. The chieftain listened, liked some verses, didn't care much for others, but invited the traders back the next evening, to give the music a second hearing. One performance led to another as the audience quickly expanded. After eight straight nights the listeners had chosen their favourite parts, and were asking the traders to sing them over and over again.

It's a shame we have no record of which ballads reached the top of the Tuzutlán charts in August 1537, for it might shed some light on why the Mayan and Catholic religions began to blend so harmoniously as soon as forced conversions were discontinued. Did the merchant-missionaries sing, for instance, of the serpent in the Garden of Eden, the treacherous, poisonous, earthbound one, without legs or wings, long associated with Mayan religious symbolism? Or did they sing of Christ's resurrection – of the man-god who flew to heaven after death, like the bloodstained Quetzal rising after Tecun Uman's defeat? The chieftain and his family, for their part, were so enthusiastic and curious about the songs that they sent an emissary back with the merchants to invite the friars to visit the highlands (apparently instructing the emissary to spy on the friars to see if they were actually as uninterested in gold, jade, and Quetzal feathers as the merchants made them out).

Receiving the emissary with joyous prayers, Las Casas's group decided to send back Father Luis Cáncer, a dedicated and well-travelled missionary who had the advantage of speaking several Indian dialects. His triumphal procession into the highlands was decorated with floral arches and celebrated with great fiestas. The *cacique* himself ordered a church to be built, and observed the first Mass. Learning of the friars' honesty, particularly of their part in Governor Maldonado's order banning Spaniards from the highlands, the chieftain converted, and urged his people to follow. He demonstrated his trust by tearing down the local idols and disavowing animal sacrifices. He was soon instrumental in arranging for the friars to extend their mission to other parts of the highlands, where this Christian minstrel show made numerous converts.

To be sure, Las Casas's success was greater among the heathens than among the Spaniards in Santiago, who were prepared to mumble 'pity' at the expected news of the meddling priest's demise, and get on with the business of subjugation. They remained unconvinced, and continued to resist Las Casas's demand that stolen native lands be returned. His practice of excommunicating Spanish slave-holders did not make Las Casas more popular. Within a few years, Las Casas's very life was threatened by the colonists. The priest had to flee to Nicaragua. Yet among the flood of official communications triggered by his humane project in the highlands, two especially indicate the brilliance of his achievement. One is a letter to the Spanish king dated November 1539, commending Las Casas for his ecclesiastical work. It was signed by Pedro de Alvarado. The second is the royal order renaming the territory 'Vera Paz' in his honour.

5

Again the flora changed completely with the elevation. No more farms. No signs of human activity at all, not even a single machine-gun nest. We were swinging up into an immense pine forest saddling craggy hills. The clouds backed off, the cicadas chorused 'cri-cri-cri-cri', and the pinelands let go a blast of sticky sweet fragrance. At 4,000 feet it was an admirable imitation of a New England summer afternoon. All that was missing was the quaint church steeple rising from the gummy air round the bend, and the Red Sox game on the radio. The nimble Toyota four-wheel became exuberant and purred through the rolling woods in fourth gear. We were running with the breeze under a pair of Swallow-tailed Kites that happened to be cruising northwards above the pine tops. Their long, white forked tails pumped slowly in the tail winds. They ignored the car, so we got to watch them for several klicks. They were in a kind of lazy hunting mode, flying smoothly and with very little effort on narrow pointed wings. Every now and then they'd make a graceful swoop down to tree line, probably to check out something on the ground that had caught their sharp eyes and might be good to eat. They had the same silhouette, two-tone colouring, and aerial ease as their smaller namesakes, the swallows. In fact, they looked like tropical swallows, swelled to twice the size of temperate-zone swallows. And they looked like they spent just about as little time on the ground. But as Hugh Land points out in *Birds of Guatemala*, they are called in Spanish *Gavilán tijereta*, which is derived from the word *tijeras*, or scissors, after the shape of their tails. Literally, scissored hawk. It seemed a more fitting name for these agile, sharp-eyed predators.

They weren't zooming around like the happy-go-lucky aerobats of Capistrano fame. They were holding something back for the attack. Even from a distance we could see the sharp little hooked beak kites use to tear and rend their meat. No swallow carried a weapon like that.

A yellow road sign stated 'Fog zone'. Precisely on cue, the fog started rolling across the mountains as we left the pinewoods behind and below. At 5,000 feet we were crossing the unmarked border into cloud forest – at last, we'd reached the habitat of the Resplendent Quetzal. A patchwork of mist and black-green peaks danced and swayed before our eyes. The road vanished around corners. It was the kind of bright, diffuse fog that might suddenly throw anything out at you – shrouded goblins, or golfers in kilts. It soothed and unnerved at the same time, snatched your bearings away, and rearranged the mountains like so many pieces of a dark puzzle. A raw, primeval, overpowering terrain. The dense wall of jungle, the towering mountains, the looming mist: all these extreme elements, combined with a sudden damp chill that drove right for your bones. You could see why the conquistadors felt superstitious and believed that mythical beasts haunted these vague ranges: diabolical water serpents and winged lizards. And you could understand, too, why it took so very long, hundreds of years after the Conquest, for the Quetzal to be transferred from the realm of the mythical to that of the known and living.

It was coming on four in the afternoon as we turned on our headlights and crept along at twenty miles an hour. The beams atomized in the vapour, spraying light uselessly out, making the road even harder to discern. I was totting up to myself all the factors that made it seem unlikely that we would see a Quetzal. To begin with, the bird is rare. No one knows whether its population numbers in the thousands or the hundreds. How could anyone know, when the army seals off entire provinces? But it's rare enough that most ornithologists believe the bird is in critical danger of becoming extinct within twenty years. And it's shy. You can't just whistle or shoosh in a Quetzal by imitating its voice: the Quetzal doesn't fall for that old birder's trick. You can't dress up like an avocado and sit in a tree till a Quetzal comes along to eat you, either. Then, also, the time of the year was wrong. Almost every study I'd read of the Quetzal had been based on field observations during the April–June breeding season, when you could at least look for a nest. Now it was mid-July. The Quetzals would probably have left their nests by now. The habitat, of course, didn't look too promising: miles and miles of fogbound jungle with tremendous canopy cover. Chasing a bird through that would be hopeless, but birding from a stand, staying in one place until a Quetzal showed up wasn't going to be much fun. At 5,500 feet, it was going to be a lot colder and wetter than I'd thought. I'd been warned. Finally, we would still need luck to elude entanglement in any military, paramilitary, or guerrilla business. From the moment we first drove up into the cloud forest, it all seemed a rather improbable folly. I could sense that our real journey was just beginning.

The immediate thing was to find shelter for the night. I was for simply pulling off the road and car-camping, crawling into our sleeping bags while they were still relatively dry. But Mickey objected strenuously. A bizarre argument broke out over our chances of getting shot.

'The guerrillas wouldn't shoot us,' he scoffed. 'They'd take the car, and the gear, and whatever money they could get. We could walk to the nearest pay phone.'

'The chances of meeting the guerrillas are pretty slim if we stay close to the road, wouldn't you say?'

'Yeah, but close to the road the chances are better of a military patrol coming by,' he said. 'Personally, I'd rather take my chances with the guerrillas than the army.'

'So the army wakes us up and we have a nice friendly chat and flash our passports at them, and they go away and leave us alone. What's the problem?'

'The problem is that they wouldn't wake us up – they'd wake us up with their Uzis.'

'What, open fire on a civilian car without any notice? That would be pretty stupid.'

'If there's one thing I've learned in Central America, it's how stupid fourteen-year-old soldiers can be!'

The fog was fast obliterating the scant daylight left us, and we both knew it was not so smart to tempt a dark, foggy night in the Guatemalan highlands. We'd hoped to locate, somewhere along this road, the Biotopo del Quetzal, but our information on the refuge was sketchy. Some said the Biotopo had fallen into disuse and neglect, maybe had even been abandoned, during recent counter-insurgency operations the army had been carrying out in the highlands. Others said the Biotopo was still an active conservation centre with a bunkhouse for itinerant naturalists. Jorge Ibarra had called it a tourist trap.

We had reached the mileage marker where the Biotopo was supposed to be, but didn't see any sign of it. A few klicks ahead we came upon the sign for the town of Purulhá, which was supposed to be in the same vicinity. Leaving the Cobán Road, we took the spur for town. The dirt road fell sharply into ruts and stones. We hadn't crawled half a mile in four-wheel drive when we were met by a civil-defence patrol manning a thick wooden pole barring the road, raised and lowered by a heavy stone weight. The Guardias consisted of four men and two boys in mufti, armed with an assortment of old single-shot rifles. They were all Indians, but they spoke Spanish.

'Halt! Halt!' They shouted as they circled the car. 'Documents!'

I handed over our passports, which one of the men backed off to study. It seemed to take him a long time. Until Mickey and I both noticed at the same time what the problem was: he was trying to read it upside down. We burst out laughing, which must have made the Indians think us slightly loco.

'Your business here?' the headman asked, returning our passports.

'We're North American scientists,' I told him to make things easy. 'We've come to look at the birds in this zone. Do you know the Biotopo del Quetzal?'

'*Ayybuen.*' He relaxed with the welcome news, which inched around the car like a humour: '*norteamericanos ... científicos ... científicos ... norteamericanos.*'

'In that case, you've gone too far,' said the Guardia. 'The Biotopo is six kilometres back.'

'The fog,' I said vaguely by way of explanation. The patrol members nodded sympathetically. I continued; 'Look, they say there might be a hotel in Purulhá, a *pensión*, somewhere we might sleep.'

They looked at each other ruefully and shook their heads. 'There is none ... There is none,' they confirmed.

'Well is there a *comedor*? A place to eat?'

'Closed ... closed,' they all agreed.

'A market maybe?'

'Thursdays and Sundays,' said the headman.

'Only?'

'Only . . . only,' they said in unison.

We decided to drive through town just for a look around, anyway. The civil guard dutifully shouldered their guns and hoisted the wooden bar, saluting the *científicos* as we passed through. We'd have done better leaving the car and going on foot. There was nothing so easily categorized as a street, only a path full of boulders and mud puddles, barely wide enough to take the little Toyota. The town consisted of several shoe stores in various states of discomposure, surrounded by dwellings made of wooden poles bound together, surmounted by steep thatched roofs. The houses fronting the path had uneven wooden boards banged together for a façade. A few were made of adobe or cinder block, eaten away by weather and time. Smaller footpaths led back among more classic Mayan wattled huts, with small, fenced-in gardens. At the far end of the village stood a concrete market shed with a corrugated metal roof, and beyond it a small adobe church, facing across a barren, treeless square, a low green concrete lodge, which was some sort of municipal building. That was it, except for the turkeys and the pigs wandering in and out of the houses as they pleased, and the wet, bedraggled mongrels blocking our way, sniffing to get our wind. There were no automobiles, no television antennas, no telephone wires, no newspaper stands: we'd effectively left the twentieth century, except for a handbill tacked outside a refreshment stand that reminded passers-by to 'Take a Coke'. The population was totally Indian, and the Indians were totally sullen. As we crept through, the women glared at us the way prisoners might stare through the bars of their cells as the warden comes by with visitors; but this was their home town, not a prison. You didn't have to be much of a mind reader to guess that they associated the arrival of strangers with trouble. Mothers dragged their children inside, out of view. There were no men in the street. There was nothing even approximating to an inn. The spur through Purulhá was only about one kilometre long. We reached the north end, and were met by a second detachment of civil guards manning a rear gate just like the front one.

'Well, so much for Purulhá,' I said. 'Looks like a fun place.'

'Oh, yes,' said the Mick. 'Outstanding hospitality!'

We turned back and headed down the Cobán Road for where the Biotopo was supposed to be.

The second time around, the Biotopo reserve had returned to its location. It was still functioning, more or less. Up a steep, short driveway we pulled on to a cobbled parking apron facing an abrupt orange dirt slope that met the forest within fifty feet. On one side were set some lean-tos and a rustic one-room wooden lodge. On the other was a mud guard-hut, a friendly riffle of smoke curling from the hole atop its thatched roof. Beyond the hut stood a thatched pergola of rough-cut branches – a kind of orientation centre, displaying placards of the three refuges in the Guatemalan system of Biotopos, each devoted to preserving the habitat of a different endangered species: the Quetzal here in the Verapaz highlands, the Manatee at Chocón-Machacas on the Caribbean Gulf and the Ocellated Turkey at Cerro Cahuí in the Petén lowlands towards Belize. So often in the American tropics, the conservation of species threatened with extinction begins and ends with protective legislation that is little enforced anyway. The fundamental discoveries of ecology, however, have taught conservationists that saving habitat must come first: without the sustenance of its environment, a species will diminish to the vanishing point no matter how well protected against poachers. Indeed, North American hunters are almost, though not entirely, right when they claim that regulated sport hunting has never driven any species into extinction. What would be more accurate to say is that hunting has a relatively minor effect on wildlife when compared to current rates of destruction of wilderness areas. But at least the immigrant peoples of the Temperate Zones of North America have a long-standing schism between their traditional romantic admiration of nature and their belief in development as the symbol of progress – the European's historical struggle to dominate his surroundings by 'improving' them. In Central America, on the contrary, a fundamental hostility has existed between the Spanish immigrant and tropical nature – man good, jungle bad, simply put. This idea runs deeply within the culture of the New World Spanish Conquest, deeper than any romantic notion of open spaces, deeper than the idea of progress. And it's not just the legacy of the conquistador's spirit of plunder either. The inter-tropical environment itself has proved a formidable, at times intractable, opponent of all human designs. The blazing heat alternating with incessant rains. The soil that looked so rich and turned out to be so poor. The feverish vegetative growth, which never gives up a season to rest, but keeps spreading, climbing, invading, throughout the year. In the Temperate Zones, if you clear some woodland, it will remain more or less clear for at least a few years. In tropical America, a cleared patch becomes, almost overnight, a thicket. Some bamboos, for example, can grow sixty feet in two months. In a year, a second-growth forest appears. This is why the machete is the Central American *campesino*'s best friend and constant companion. The hyper-growth of the tropics can seem like a nasty obstruction. I suppose it's hard to love land you never actually get to see. The Biotopo system depicted on the placards looked like a step in the right direction. I had to marvel that someone had managed to bring it off in a country as deeply at war with itself as Guatemala.

Under the thatched pergola in the waning light stood a fine-boned, dark-skinned man in a grey ranger's uniform, accompanied by a little boy who came up to his mid-thigh. This was the Biotopo's lone ranger, Fernando Ack, and his angelic son Francisco, who were soon to become our guides and friends. Fernando was talking with two paunchy, white Ladino men dressed in creased slacks and short-sleeved sports shirts and white patent leather loafers: the looked as though they'd got lost on a junket to play the slots in Atlantic City. We sat down on the wooden bench to wait, and eavesdropped on the conversation. The two Ladinos were talking down to the Indian park ranger, asking a lot of insolent and stupid questions:

'The Quetzal – you got 'em here, or what?'

'So where are they?'

'The weather always like this up here?'

'What, you live in that dirt hut there?'

'Show us the Quetzals, we gotta go. We drove all the way from [unintelligible].'

Fernando calmly fielded their questions in his appealing and gentle manner. His voice floated through the saturated air like a flute wandering over a soft melody. Fernando, too, punctuated his speech with the expletive we'd heard earlier at the Purulhá gate, '*Ayyybuen*,' slowly ascending the scale on the first syllable, then dropping back to earth on the second. His fluent Spanish seemed to carry the whole flavour of the Mayan highlands. To everything Fernando said, little Francisco nodded grave assent. He stood by his father like a faithful pup, the knees of his red pants gone. He hung on his dad's every word, and soon reached around Fernando's leg to anchor himself more firmly. It was difficult to tell who was protecting whom. Then he threw his head all the way back, the better to gaze up at the tall adults, as though they were trees. His head rolled from one side to the other as he followed their conversation – when, suddenly, he seemed to discover the dizzying thrill of that motion. He giggled wildly, and clamped his hand over his mouth to stifle its escape. Without pausing, Fernando mildly rattled his leg, and stroked his son's head.

The conversation droned on for a while longer. The two men couldn't decide whether they should chance walking up one of the easier trails in the half hour or so left before dark. It was foggy and damp, and their clothes weren't any good for hiking. They said they hadn't realized that you had to get out of the car to see a Quetzal. Fernando nodded sympathetically. Finally, one of them asked if there were any snakes around. Fernando matter-of-factly replied that a fer-de-lance – that deadly Central American serpent – lived in the undergrowth just above the orange slope. That did it. The city slickers turned tail without so much as a thank you and elbowed each other in their rush to get back to their car. Fernando watched them leave with an unruffled, oriental expression of contemplation. Francisco stole a bemused glance at us, and shrugged his shoulders.

I asked, 'Is there really a fer-de-lance?'

'*Claro que sí,*' affirmed Fernando. 'But the chances of a confrontation are not great. I've only seen the snake twice in all the time I've worked here. It stays in the undergrowth above the camping area. I think it waits for rats that come to feed on the garbage. It's a very shy serpent.'

We introduced ourselves and explained our purpose, as well as our immediate predicament of having no roof for the night. Fernando was most hospitable. 'You can stay here,' he said, 'in your car or in the shelters, as you wish. Stay as long as you like, you'll be safe. I'm here twenty-four hours a day, making my rounds or in the guardhouse. Unfortunately, the other park rangers are' – he hesitated to find the right word – 'on vacation. So right now I'm alone.'

At this Francisco's face fell into a grey sketch, and Fernando quickly corrected himself: 'We're here alone, that is, my son Francisco and I.'

We moved down to the apron of cobbles, and watched the weather roll over the forested ridges below. The mist was magnificent, now that we didn't have to travel in it, more secretive and smokier than I had noticed. It broke only to cast up stark black-and-white images: the lone skeleton of a dead tree; a waterfall perched in clouds; two grey humps, smooth, round, and distant, like a pair of surfacing whales. Blustery clouds, shadowed around the edges, were dissolving into local rain showers, which the highland winds swept aslant. Over all this we could hear a tripping, trilling bird song tumbling down the valleys – dah-dah-dah-chirkee-werkee, cher-che-wer-but-but-reeba-chip – like a stone slipping helplessly away in a rushing torrent.

'Listen,' said Francisco, 'that's the *guardabarranca*.'

It was the guardian of the valleys: the Brown-backed Solitaire.

Had they observed any Quetzals lately, I asked Fernando eagerly.

'It's varied on account of the weather,' he replied. 'This time of year the Quetzals sometimes come down from the greater heights to find fruits to eat down here at more modest elevations.' He pointed out some broad-leaved *aguacatillo* saplings planted near the parking area. 'We started planting these a few years ago when the Biotopo opened, to attract the birds within sight of visitors. The birds do come to the trees, to feed on the small fruits. But I'm not sure how good an idea it is. If too many people drive here, jump out of their cars, and see the Quetzals, they may go home thinking, If it's so easy to find them, there must be plenty left, so why worry about them, they can't be in any danger.'

On the other hand, he continued, the Biotopo seemed to be succeeding in stabilizing or even increasing the local Quetzal population. Students from the University of San Carlos, who had carried out a breeding census, had estimated that fifty nesting pairs of Quetzals were now residing within the Biotopo's approximately 4,500 acres, and nearby environs. 'It's getting crowded in our little ark,' Fernando said with a broad gesture across the Biotopo domain. 'We're running out of dead tree trunks for nest holes. If you had come last year, you would have seen Quetzals nesting just across this first *barranca*, in that clearing where the trunk is. You could stand right here and watch them. When the first pair left the nest hole, a second pair moved in.'

'Do they repeat at the same nest site year after year?'

'We don't think so,' said Fernando. 'It seems that widening a woodpecker's hole is part of their courtship ritual.'

'Have you tried building artificial nest sites?'

'A student from the university put up some the past few years. But so far the birds haven't used them.'

Fernando was so knowledgeable and soft-spoken I could have gone on listening to him for hours. But in the midst of this briefing, Francisco, who had been doing his characteristic grave nod, suddenly jumped up and started tugging insistently at his father's sleeve.

'*Allí está,*' he whispered excitedly. 'There it is.'

'What?' I asked him.

'*El Quetzal!*' His brown eyes burned with magical intensity. 'Listen.'

He was responding to a high-pitched call of two notes, a hollow 'whoop, whoop', similar in timbre to the voice of a male cardinal.

Off we ran, following the boy's lead. The trail was graded in switchbacks and well maintained. Francisco strode ahead with the utter self-assurance of a relief pitcher coming in with bases loaded in the ninth. His little fists were pumping before his chest in the syncopated rhythm I'd noticed the Indian women using as they walked the roads. He didn't have to keep his eyes down where his rubber boots were treading, but gazed around alertly, eyes and ears poised in expectation. It was harder for me. At ground level, the fog and dense foliage operated like heavy curtains. The earth beneath our feet squished like a sponge – I wanted to be careful not to step on anything that bites. Through crowded vines and epiphytes, the canopy overhead was too thick even to identify the leaves from the bark. It was only a shade or two shy of nightfall. I couldn't have found a bird up there with radar. It took the adept ten-year-old all of about thirty seconds. He gauged the direction and distance of the Quetzal's call, sized up the location where it must be, moved in on the particular tree, and – bingo! – Francisco had spotted a Quetzal. Then he casually called us over, pointed into the canopy, to a tree behind a tree behind another tree, and trotted back beside me for a pat on the head. I thought: that Osbert Salvin must have been one hell of a shot.

For a long while I continued to see *nada* in the canopy, except undifferentiated tree-tops. Fernando patiently whispered directions, however, and finally I did sight my first Q-bird in the over-arching foliage – or, rather, I should say, my first Quetzal *part*, since all I could really make out was an oblong red splotch, no head or wings, and certainly no majestic tail plumes. The bird, if it was the bird, perched calmly for a few minutes (Quetzals are notable 'sitters', so this increased my will to believe that the red splotch actually belonged to one). It gave me time to move around in the ferns until I got into the best possible line of sight. This barely enabled me to observe the outline of a head and back. The bird looked azure in the grey dusk.

'Is it a male?' I asked Fernando.

'It could be either,' he said. 'It's impossible to say at such a distance, in this light. The *hembra* [hen] lacks the long tail plumes and head crest of the *macho*, but is otherwise very similar.'

Since neither tail nor head crest were visible, I couldn't tell whether this was the *hembra* or the *macho* Quetzal. Truth be told, I wouldn't have known it was a Quetzal at all if it weren't for the little boy's natural skills as a bird finder. Only one of those crazed 'listers', who run around the world checking off new species of birds they see, could have taken the least satisfaction in such an ambiguous sighting. Yet it did establish that there were Quetzals in the vicinity – and thus, that we would have to travel no farther for the time being. Finally, the oblong red splotch in the canopy disappeared. I said, 'Good work, Francisco. You come with us tomorrow and be our eyes.'

He squealed with delight, and asked his amused father if he could. Fernando asked, 'Then you'll be staying with us for a while?'

It felt good to have arrived somewhere.

6

After dark the rain began in earnest, and then the convoys – a steady stream of military trucks rumbling north up the Cobán Road all through the night. Their head beams sprayed fog like the watchtower spotlights in an old prison escape movie, and their huge shadows climbed the wall of fog created by the headlights of the trucks coming up behind. Even the Guatemalan Army can look one hundred feet tall with the right special effects going for it. I rolled up the car windows against the noise and rain, but it got unbearably muggy inside the Toyota. Mistake number one: in my elation over having found a likely spot for Quetzal watching, I celebrated with a few shots of rotgut offered by two campers at the Biotopo whom we'd run into after dark – completely forgetting that we hadn't eaten all day. The campers had a small fire going in one of the lean-tos. When they invited us over to share it, I thought I recognized one of them. She turned out to be Alfredo Schlehauf's niece Gisela, the young woman visiting from Germany. The young man with her was Schlehauf's nephew William from Cobán, and he had the bottle. He called it Guatemalan 'vodka'.

'Have you gone birding up at Finca Remedios?' I asked.

'Oh, no,' said Gisela Schlehauf. 'My uncle won't let us visit the farm at all. It's too dangerous there.'

'This is safe?'

William Schlehauf said, 'I used to work with my uncle at the *finca*, but now I don't go there either. The guerrillas have been burning farms in the area. You won't be visiting at the best time for my uncle. Actually, we're practically out of business. Have another drink?'

Afterwards, the raw liquor took its predictable toll. Dizzy and nauseous by turns, I'd lean out of the door every so often and throw up, until my stomach emptied itself of the poison. Then I flopped back naked outside my sleeping bag, listening to the convoy trucks smacking a pothole – 'ping-*ponk*, ping-*ponk*!' It was like counting steel sheep. In the distance, I could hear an explosion; it sounded like a mortar going off. You didn't have to be drunk to sense the terror of the Indian families living up in the hillside hamlets as they listened to the convoys rumbling through the rain, but it helped focus the mind. I listened for the screech of brakes, the boots sloshing in mud, the slamming fists on the door, the staccato bursts of automatic weapons.

'Aw, Christ, knock it off,' said Mickey, who had of course abstained, and was resting easy. 'Will you let me sleep?'

'That wasn't me, you idiot,' I groaned.

'You poor guy,' he taunted. 'You better get some shut-eye. We have to be up in a few hours to find the Quetzal.'

I swore something. Then I threw my boot at him.

He sat up and heard the trucks. 'Convoys,' he said. 'Too bad we're not camping on the road like you suggested. I could get excellent pictures of night operations.'

I was too out of it to respond. But I deserved the butt end of his humour: I shuddered to even think of the living nightmare that might have taken place if we'd car-camped on the road. But then my stomach muscles seized up again, I leaned out and dry-heaved, and thought of nothing except a resolution never to drink again on an empty stomach.

'Poor guy,' my partner repeated, adding mystifyingly, 'stop strangling that snake!' Then he turned over and went right back to sleep.

7

In the morning we went out for a long hike into the Biotopo reserve, not so much thinking to repeat the freak luck of seeing Quetzals as hoping to shake off the road dust and recon the cloud forest. It was difficult at first, for one unaccustomed to elevated montane tropical forest, to know whether it was actually raining from the sky above the canopy, or simply dripping from the trees. Tiny torrents trickled down the trunks, fat globules lined up in ranks on leaves, and miniature showers shook down from the upper foliage, at times with such force, but with such narrow scope, that you felt sure there was a monkey up there playing a prank on the interlopers below. But as many times as I lifted my eyes during the course of the day, only one monkey came into view – a black Howler Monkey placidly browsing in the tree-tops. It was neither howling nor playing a practical joke. In fact, the howler seemed remote and indifferent to the forest floor one hundred feet below: it crept through the trees at the arthritic pace of a patriarch. An elderly male, by all odds.

I was astonished at what a watery place this was. The trade-winds driven by the earth's rotation suck up moisture crossing the Caribbean Sea. When they meet the mountainous backbone of the Central American isthmus, the humid air rises, cools, and deposits its load in almost every form of precipitation, including, at these elevations, frosty morning dews and occasional snow showers. But mostly in the form of rain – sometimes 315 days a year of it. It's this nearly constant supply of moisture that largely accounts for the luxuriant diversity of the cloud forest's flora. Water is the lifeblood of what has been called the most complex ecosystem on earth. Wandering from 3,000 to 6,000 feet elevation, one noticed repeatedly how the Central American climate and geography were conducive to the cohabitation of purely tropical plants and more familiar Temperate-Zone forms. Only a few hundred feet would mark the distance between gigantic hairy tree ferns and the familiar oaks of North America. On a sunny, steaming incline, epiphyte orchids festooned the trees, while homely pines jutted from a nearby shaded nook.

What a fruitful meeting of north and south. What a nice tolerance plants of such different origins had for each other, even though they competed for space and nutrients. Here were the species that had been moving back and forth in that 'sweepstakes' pattern, ever since a continuous terrestrial connection between the northern and southern hemispheres had begun to form in the Paleozoic era. Their seed didn't need passports, visas, safe conducts. They just got into the belly of a bird and rode, or flew, on the wind. Botanically, at least, Central America seemed eminently free.

Distinct from the poor red laterite soils we had seen on our journey, the floor of the cloud forest was a deep loamy brown, mossy and springy underfoot. The air smelt strongly of the wet humus, and I thought: if I had weather conditions like this in my New Jersey garden, my cucumbers might look like the Goodyear blimp. As this was mid-July, we were in the very middle of the heaviest season of rainfall. Central Americans call it their *invierno*, or winter. Of course it's not really a true winter. The trees and shrubs don't go through a dormant period. They just call it winter, I suppose, to have what the gringos have in terms of seasons. Running from mid-May until mid-October, the wet season is marked by, if not constant, then at least daily precipitation, often combined with the kind of sock-eyed fog we'd experienced the previous evening. It's also the season of mud slides, rock falls, road washouts, floods, and other local calamities that turn people to superstition, not without cause. We had already heard the Guatemalan belief that heavy inundations foretell earthquakes. But whether based on common sense and observation, or on the infamous coincidence of rains and the volcanic eruption and earthquake that

destroyed Ciudad Vieja under Alvarado's unlucky widow Doña
Beatriz, I don't know. We had also heard of a short dry spell at this time
of year, known as the *canícula*, a week or ten days of sunny, spring-
like weather wedged between the initial rainy season and the later
September–October phase, with its danger of hurricanes. But who to
apply to for the advent of this 'little summer'?

Although neo-tropical birds vary radically in their life cycles, many
species of the montane forest, including the Quetzal, build their nests
and bear their young during the initial months of the rainy season,
when the increased precipitation spurs fruiting for feeding nestlings,
and perhaps also when migratory birds, potentially competitive for
food and nesting territories, have flown north to breed in the Temperate Zone summer: as is well known, year-round residents have their
best juice after the seasonal crowds go home. All thirty-four species of
the New World trogons, the family to which the Quetzal belongs,
seem to breed during this wet season. At its height, the forest is alive
with territorial calls and mating songs. But by now, with abandoned
nests and cold drizzles, most of the birds were lying low. Occasionally,
a hummingbird zipped by, undaunted in its hunt for nectar and insects
among the low-growing tubular red blooms of the *alcants*. The occasional fluty tones of Grey Silky Flycatchers drifted up the ridges. Several
times I heard birds hopping through the leaves, or caught an evanes-
cent flash of wings. But the birds blended perfectly into the verdure.
Chasing them into the matted undergrowth was out of the question,
because wherever the trail had been cut through the forest, the second
growth had spurted up thickly. The deeper forest, where the shrub
layer was thinner and the view more open, was in the birding dol-
drums.

In any case, there would be no need to worry about bathing. The
floor of the cloud forest has been described as an immense sponge.
The level above ground could be equally well described as a kind
of continually running shower stall. Resisting the wash at first like
recalcitrant cats facing the flea tub, we soon gave in to the inevitable
soaking, then began to shed our clothes as the temperature rose and
the humidity increased. Eventually, we came to an icy swift stream,
where pools had been deepened for swimming. The temptation was
too much. We threw off our clothes and took a bracing plunge in
the company of small lizards sunning themselves on stepping-stones.

It's amazing what a swim in a mountain stream can do for your spirits.

For the first time since coming to Guatemala, I felt a shifting of gears. The accumulated tensions of the city and road washed downstream like so many air bubbles, and we were left squeaky clean to admire the peculiar monochrome beauty of the glossy green jungle. The shaded gloom of the canopy was cut here and there by sharp beams of sun. Wherever a shaft penetrated a gap in the upper storey, a packed garden of seedlings had formed, thrusting their heads up and their roots down in a botanical Olympics to receive more light, and clutch more nutrients, than competitors hard by. The woody creepers seemed to have much the upper hand in this competition. Strangler figs swarmed round thick tree trunks, and liana vines snaked through the lower storey, seeking something to cling to, wantonly constricting anything in their path. Their unbridled longitudinal growth gave the forest a sense of movement and activity not usually associated with plants, which we tend to think of as rather stationary, passive creatures. But they're not. Plants have very distinct personalities. The vines are the hustlers, the over-achievers of the tropical forest. The profuse epiphytes and bromeliads, surviving only on what can be gleaned from the moist air, are more like innocent children: even their unkempt, dangling air roots gave the impression of playful plenitude.

The cloud forest is one place where all the capacities of language, art, and even natural science are simply not up to the job of conveying the bewildering complexity, the infinite variety, and the daunting sense of how many of tropical nature's ways are still beyond our ken. Relations of predator and prey both delicate and savage. Food webs both simple and enormously complex. Private harmonies of insects with plants, birds with plants. Beetles that spend their entire lives hidden under one leaf, and moths that spend their entire lives mimicking another species. Whole communities that we barely know exist, and plants no human has ever seen or put a name on. The jungle keeps its best secrets to itself. The whole is a lot greater than the sum of its parts. As the young Darwin wrote in his notebook when he came to the American tropics on HMS *Beagle*: 'The form of the orange tree, the cocoa-nut, the palm, the mango, the treefern, the banana, will remain clear and separate: but the thousand beauties which unite these into one perfect scene must fade away; yet they will leave, like a tale heard in childhood, a picture full of indistinct, but most beautiful, figures.'

It's said that only those who have entered a tropical rain forest will ever know what a rain forest is like, or have any inkling of its significance in nature. I believe it. And also that only those who spend a lifetime studying the tropical forests will understand how ignorant we remain of its inner workings. It was by such thoughts, lying in a speck of sunlight drying off from our brisk bath, that I was reminded that we had entered no vast expanse of uncharted territory, but only a single reserve of several thousand acres. A few kilometres below ran the Cobán Road, with its ugly troop movements, its semis fuming diesel exhaust. Down the Biotopo's western ridge, in the valley of Salamá, the Guatemalan government has been building an extensive hydroelectric project. This self-conscious patch of cloud forest that has been preserved is exactly that: a remnant – in a sense, a legend of what the extensive cloud forests of Central America were in a former time, before Western man obtained the capacity or felt the need to obliterate them.

8

It was Fernando who told me there was another gringo in the area. *El Voluntario*, the local people called him: The Volunteer. We ran into him at dusk, wandering in the fog near the Biotopo parking lot, looking lost and pale. He was a Peace Corps volunteer, and he'd come out to the highlands six weeks ago to do a natural resources survey of the Biotopo area, but he hadn't started yet. One problem was that no one in authority had been informed he was coming, if there was anyone in authority. And since he'd failed his training course in Spanish, he couldn't make himself understood so well. In addition, he'd been wasted by virulent dysentery ever since his first meal in Purulhá, and he'd spent a lot of his first six weeks there running through the rain to the wood-slat privy out behind the house he'd rented, or lying sick on his bunk. He was ashen and dehydrated, and said, 'Gosh, am I glad to see you guys. Bring any toilet paper with you, by any chance?'

We told him we'd come to watch Quetzals, and that we probably had a spare roll in the back of the Toyota.

The Volunteer invited us over to his house. It stood across a muddy road from a low-lying pasture with a swollen stream running through it, about half a mile from the town. It had a sturdy metal roof and a concrete floor. I told him we'd give him all the toilet paper he needed, if we could sleep there. He agreed. The Volunteer was an earnest twenty-five-year-old Midwesterner, gawky, shy, and fresh out of college. He'd never been out of the Midwest before, and didn't know what to make of Central America. 'It's sure not like home,' he said wistfully. He was right there.

Consequently, he was homesick. Lying around with a stomach-ache and the runs didn't improve his morale. All the next day, and the one after that, the rains drummed against the corrugated metal roof, and the Volunteer was the only bird we got to watch. The townspeople kept coming to the Volunteer's house. The kids walked into his yard and took water from his spigot. The town elders came and told him they needed a public bus to get their produce to market. The Indian women asked freely for food and money. They thought a gringo might have some influence with the authorities, or at least provide some protection. The Volunteer smiled his agreeable Midwest grin, but he was clearly overwhelmed, and hadn't the slightest idea what to do. 'I'm not really a political person,' he reasoned with us over and over as the rain slashed through the windows, which had no panes. 'The Peace Corps told us to just do our jobs and stay out of things. I'm a resource biologist, not an activist.'

'What the hell's the difference?' Mickey asked.

The Volunteer thought he knew before coming out to the highlands, but now he wasn't so sure. He'd been assigned to work with INGUAT, the Guatemalan tourist organization located in Guatemala City. They told him to make a biological survey of the lands on the perimeter of the Biotopo, which later might be purchased to extend the reserve. A worthy task, on the face of it. But aside from the fact that you'd really have to be a hotshot tropical ecologist to single-handedly survey the plants and animals of such a large territory, the

Volunteer required written permissions just to go on to privately owned lands. No one locally knew who the lands belonged to. Neither did the INGUAT officials he'd met with in Guatemala City. The INGUAT officials said they couldn't help him. They were too pre-occupied trying to promote tourism in the country Amnesty International was calling the worst human rights violator in the Western hemisphere, a thankless as well as hopeless job. So what was the Volunteer doing in Purulhá? Not much. Running to the john. Trying to get his digestive system under control. Dealing himself hands of patience as the rains hardened into inundations. And, undoubtedly, beginning to think of chucking it all and going home. He spoke of cases where the Peace Corps had sent volunteers back to the States, if their dysentery got bad enough. Every meal of *tortillas*, beans, and chillies, in the local *comedor* was another step in that direction.

'Sad Sack,' Mickey dubbed him.

Nevertheless, we were glad not to have to sleep in the car, as the squalls marched across the ridges night and day. Besides, by staying in Purulhá, we were able to meet some of the local characters, and make at least one friend in the rural town. This was the congenial *mestizo* Don Lázaro, a lineman for the national electric company, who lived just up the hill behind the Volunteer's concrete bungalow, and was actually his landlord. Don Lázaro showed us around his yard, where, until recently, he'd had several Quetzals visiting daily to find food. Formerly the supervisor of trail construction at the Biotopo, Don Lázaro and his extended family occupied a rickety and rambling old wooden house filled with kids and women. He came out in shades and a hard hat and introduced us to his wife, a small, darkly beautiful Indian-featured woman wearing her black hair in long braids. Their front yard was filled with a fine collection of wild orchids and wild herbs they'd transplanted from the forest. Above stood several varieties of fruit trees – citrus, papaya, sapodilla – enough to keep the family supplied, as well as several *aguacatillo* trees, the same miniature avocados planted at the Biotopo to attract the fruit-eating Quetzals. Don Lázaro had planted them there for that very purpose – or more specifically, so that his children could watch the hemisphere's most beautiful bird from the front porch.

'Oh, yes, the Quetzals love those *aguacatillos*,' he said. 'And the sapodillas, too.' He showed us the fruits of the *aguacatillo*, slightly larger than a cherry, hard as a crab-apple, and pale lime green in colour. Don Lázaro then gathered a few of the fallen, unripe fruits of the sapodilla, the chicle trees from which the latex for chewing gum is extracted. The dark leathery leaves of the sapodilla were like those of the related rubber tree, though not as wide or thick, and the sapodilla fruits were covered with a coarse, wrinkled brown skin. They smelled very much like almonds. Don Lázaro spoke of the birds with authority and affection. 'Five years ago, we had six Quetzals here, three pairs, feeding in the fruit trees. They were *bastante meticulosas* – rather meticulous – in the feeding hours they kept. They would appear each morning promptly at six, and remain near the yard, feeding off and on, until seven-thirty, or perhaps eight. Then they returned again at four in the afternoon, and stayed until five, rain or shine. The children would sit on the porch and watch them.'

'It was wonderful for the little ones,' added his wife. 'The birds were so entertaining.'

Don Lázaro went on: 'By last October we were down to one single *hembra*. And October was the last time we saw a Quetzal in Purulhá. Afterwards? I don't know – they became disappeared.'

I glanced at Don Lázaro: he had used the word *desaparecido* – disappeared – a word commonly used in Central America to refer to people, not birds, and one not used with impunity in the presence of strangers in Guatemala. But I couldn't discern any irony in his face or voice. He added, 'I can't say whether the Quetzals left the area, or perhaps were killed.'

I asked if the Quetzals might have ranged into the friendly confines of the Biotopo.

'Sure, could be,' he answered. 'Out here they're in danger. The Indians who live up in the mountains are *muy primitivo*, very primitive. They kill the Quetzal for food. I've seen this with my own eyes, in the villages beyond Chicoi. The Indians don't know the Quetzal is the symbol of the nation. They are illiterate and uneducated. They call the bird *pajarito*, and think it's some kind of parrot. They say the Quetzal makes a tasty stew. Then again, at times the boys around town hit birds with their slingshots. The Quetzals are so large and bright. They make an easy target.'

9

At Don Lázaro's suggestion we went up into the hills north and east of Purulhá, heading towards the local arch-aeological site he'd mentioned called Chicoi. The passable road ended abruptly at the gate of an extensive *finca*, which spread across a broad, soggy meadow. A thick carpet of hyacinths bloomed violet there. In its perfumed midst sat the unaccountable hulk of a twin-engine plane, post-war vintage. Woollen sheep placidly grazed the wet pasture. Their chocolate and vanilla coats provided points of reference as the gentle contours of the valley slipped away into blue-tinted foothills. Even our Sad Sack Volunteer, who'd tagged along in the hope of seeing a Quetzal, seemed to cheer up at the gush of hot sunshine unexpectedly falling out of the sky. If the 'little summer' lasted only a day, or an afternoon, it would be worth these few sparkling hours, combining the gusty breezes of April mornings with the piercing rain of August.

It was only half an hour's walk to the draw of the valley, where we hesitated, unsure of the trail and wary of trespassing. Luckily, a grisly old Indian *peón* emerged from the maize hedge surrounding his wattled hut, and said in Spanish, 'Ah, you must be looking for *La Cueva* – the cave. It's more up.' He barked something in Kekchí at his wife, who was sitting on a rock in their yard sewing, a child playing at her feet, and without waiting for a response, led us uphill himself. We crossed a swift-flowing stream, swollen with mud by the rains, over the broken, mossy stones of an ancient bridge.

'*El río*,' said the Old Man, not without a touch of local pride, but it was really just a winding creek, not worthy of the title river, nor worthy of a stone bridge – which was itself no longer worthy of a real river in any case. We came to a second gate in a barbed-wire fence, marking the rear limit of the *finca* we had already dubbed 'Twin Engines' in honour of the woebegone plane. Whistling under his breath, the old man herded the sheep away from the low-growing, thorny vines they were browsing for an unfamiliar tiny yellow fruit, like a miniature squash. 'Don't eat that,' our elderly friend advised us in all seriousness. 'The sheep get drunk when they eat it. But if a man eats it, your head leaves your body for another place and you die. We call it *espinas* – thistles.'

'We won't eat the thistles,' I promised him.

He showed us where to pass under a hole in the barbed wire and bade us the traditional goodbye: 'That it may go well with you,' adding, 'Remember, more up. Just ask for the cave. Everyone knows it.'

'More up' the steep and sun-drenched southern slope it was like entering another world: the world of the pre-Columbian farmer turned vertical against the side of the mountain, as if wrenched upright by a millennium of earthquakes. All around us spread the Indian *milpas*, much as they would have appeared several thousand years ago, when the people of the New World were first bringing under cultivation the wild-grass ancestor of corn, just as the peoples of the Old World domesticated their native grasses of wheat, barley, rye, oats, and millet. Except, as I say, skewed upright, but that angle made all the difference. The nine-month (May–January) growing season was less than a third gone, and the ochre-coloured earth between the green maize sprouts was badly scored and gullied. Each section, separated invisibly to the ignorant observer, was being tended in the high afternoon by a grown man and one or two sons. Some were cultivating their maize with heavy-handed hoes, others clearing off burnt tree stumps and live scrub with their machetes. The tools clanged and echoed in the thin air, and the bundles of fuel wood were loaded on to the backs of the children, who threaded their way down into the valley under their inordinate burdens. The higher we climbed, the more the mountain sides looked as though a ferocious behemoth had clawed them blindly. At about 4,500 feet you could turn round and view the panorama, resembling any of the thousands of such highlands tableaux: the treeless soil washing away under the rains, the flooding meadows, which would be bone dry come January, the fifty or seventy-five Indians tending their isolated, eroded little plots on the incline, while domestic livestock ranged the fertile valley, and the blue-and-white Guatemalan flags sticking up out of the *milpas*, which the general had cynically ordered the population to raise in a show of support for his regime. The farmers didn't dare not fly one. Suddenly all sophisticated arguments about democracy and totalitarianism, capitalism and communism, unravelled into meaningless abstraction before the highlands landscape. Erosion, population, land hunger, water – these seemed the only subjects worthy of reflection in a country where the Ladinos long ago appropriated the valleys for cash crops and pasture, where the Indians were left to scrape away at the hillside forests with Iron Age implements.

Only in one spot had a curious fringe of primary forest been left standing. It rose as a clump of dense, tall trees, like a patch of thick green hair protruding from a scalped head: 'La Cueva?' I shouted my question to a farmer two football fields above us. He signalled back with a wide sweep of his arm towards the stand of greenery. And there, peering through prodigious thickets, you lost your breath: great shafts of sunlight slanted down through the trees, then down and down a deep pit, 200 feet down to the bottom, where the cave of Chicoi gaped open like an onyx mouth. The sunbeams were broken by moisture and specks of dust into flashes and bands of incandescent colours – blue-green, green-gold, and golden. And all the way down that immense shimmering hole in the earth's surface, as though a plug had been lifted, stretched the arched black cave. I hadn't expected anything so monumental, so ancient, so mystical. We had to climb down and take a look.

Chicoi wasn't the type of site to excite an ambitious archaeologist. You could tell from a glance there wasn't going to be any exquisite Mayan temple buried under all that grey-and-white sediment, only perhaps some bones, the remains of meals and fires, petrified seeds, shards of pottery – all very likely disarranged by successive residents. It didn't look at all like the right place to discover the key to Mayan writing, or any clues about the infamous and still unexplained collapse of the Classic Mayan civilization around AD 1000. Chicoi probably antedated all that: only the early cave dwellers would have sought the protection of such gloomy digs. Yet as such, Chicoi didn't need temples, tablets, or potsherds. The image of human beginnings in the Americas was all there, written into the topography: the long migrations from Asia, the eastward course of peoples towards the rising sun, which some Mayanists see as the original source of the sun worship common to so many Meso-American religions. The sun worship itself. The hunting-gathering of the first settlers living in caves, and their eventual emergence above ground as maize was brought under cultivation.

Along the pitched trail winding down into the hole like rifling, this impression of consanguinity was reinforced. Below the surface, the sun's rays fell with an intense clarity and direction, pinpointing the vegetation cropping out of the vertical rock face. As we descended farther, the music of water drops began, a wooden plunking as they hit the ground outside the cave door. At the bottom, the cave mouth formed an awesome archway, but the cave behind it was rather shallow, like a great stone stage. Stout vines and limestone stalactites dangled down from the milky-coloured roof. At the rear wall, among the stone and wood rubble on the floor, someone had erected a shrine of three burning candles; the feeble glow flickering across the cave wall had been completely hidden from view above. But there was no doubt that Chicoi was still in use as a shrine, altar, or place of worship. The Indians had taken their religion back down to the cave of their forefathers where it came from.

I wondered what the general could do about that.

10

In Mayan times, so Las Casas tells us, it was no joking matter to kill a Quetzal. 'In the province of Vera Paz they punish with death him who killed the bird with the rich plumes because it is not found in other places, and these feathers were things of great value because they used them as money.' Other Spanish chroniclers of the post-Conquest period use the phrase, 'they preferred them to gold', which may be more accurate. Either way, the lavish green plumes of the male Quetzal were a precious trade item, carried from the Guatemalan highlands to the great lowland Mayan religious centres of Chichén Itzá, Uxmal, and Copán, where the cult of Quetzalcoatl flourished in the pre-Columbian era. The carriers of these cultural treasures were itinerant vendors known as *pochtecas* – literally 'merchants who lead'. They ranged out from Cholula, a city far north on the Mexican plateau, to ply their wares across the Yucatán peninsula, Mayan Central America, and beyond. The geographical range of the *pochtecas*' trade routes was apparently limited only by their search for 'The Land of the Sun', the paradise Quetzalcoatl had set sail for when he left Cholula on his heroic journey. Wherever evidence of the green Quetzal plume is found in art or writing, the likelihood exists of one of these merchants having anciently visited – and archaeological evidence of Quetzal plumes has been found as far north as New Mexico, where the Zuni Indians used them for ceremonial purposes, and as far south as the Andes, among remnant tribes of the Inca Empire.

Although the *pochtecas* were apparently hard-headed businessmen, members of their guild were explicitly forbidden from accumulating individual wealth or displaying personal affluence. What kind of goods they dealt in, Mayanists can only speculate, but the *pochtecas*' motive was not big sales. Nor was profit their end. They were sworn to act with the humility and modesty of persons befitting a religious order. Theirs was a brotherhood of mystical travelling salesmen, more devoted to spreading the cult of Quetzalcoatl than to the mere buying and selling of prized plumes, precious jade, or broom handles, for that matter.

Not very much is known about the *pochtecas*, but they seem to have acted as advisers to tribal chieftains, or as roving ambassadors. They travelled with impunity between warring tribes, and were evidently extended the hospitality befitting a privileged class. The *pochtecas* supplied the native nobility of many tribes with the Quetzal feathers that were fashioned into elaborate headdresses, and worn as symbols of authority during ceremonial occasions. It's far from improbable that on the buying end of their travels, the *pochtecas* taught the highland peoples of southern Mexico, Guatemala, Honduras, and Nicaragua to remove the tail feathers from the male Quetzal and release it unharmed, according to the following method recorded by a Dominican priest called Hernández in 1651:

> The fowlers betake themselves to the Mountains, and there hiding themselves in small cottages, scatter up and down boil'd Indian Wheat, and prick down in the ground many rods besmeared with Bird-lime, wherewith the Birds intangled become their prey.

The *pochtecas* must also have known a good deal about the

Quetzal's nesting habits and habitat requirements, because there is no evidence whatsoever of the natives trying to keep Quetzals in captivity. Furthermore, as dispensers of Quetzalcoatl's laws, they may well have been the ones to institute bans on the killing of Quetzals. Capital punishment for killing a bird may sound stiff, but this native conservation measure was of such emotional power that it was still in force when Las Casas came to the highlands, 600 years after the collapse of Classic Mayan civilization.

As Quetzalcoatl was the *pochtecas'* protector and patron, so they dedicated themselves to the religious principles of this demigod, endlessly portrayed in pre-Columbian art as the plumed serpent. Who was Quetzalcoatl? What exactly were the principles of his cult? Unfortunately, the answers remain shrouded, distant, mysterious, and slurred by time. The image of the plumed serpent appears with stupendous consistency over a vast range of different native Meso-American cultures and epochs, going back some three thousand years. This amalgam of bird and snake, flyer and crawler, fruit eater and carnivore, gentleness and violence, is found among the Toltecs, Olmecs, Mixtecs, Zapotecs, and Aztecs, as well as the Mayans. But most of what might have helped scholars interpret the plumed serpent figure went up in smoke when the Spanish burned the native holy books. They were heretical, of course. There is no way of estimating how many of these written native mythologies were destroyed: in Bishop de Landa's infamous *auto-da-fé* of July 1562, no fewer than twenty-seven of the treasured texts were consigned to the flames in a single outing. Who knows how many dozens or hundreds more were lost on the run, buried in caves, committed to memory and hastily hidden in the jungle? As a result, what knowledge remains of the culture hero Quetzalcoatl has been culled from a few painted murals on temple walls; from the two scrolls of Mayan writing that survived the Conquest more or less intact; and from the shards of earlier myths, rewritten (and very likely distorted) in several post-Conquest native chronicles.

From these sources, nevertheless, it seems clear that there was a historical King Quetzalcoatl, remembered as the great lawgiver and civilizer, inventor of the calendar, as well as of medicine, and the leader under whom maize was domesticated as the staple of the human diet. The French Mayanist Séjourné believed that such a monarch lived around the dawn of the Common Era. Suggestions that he was actually one of Jesus' disciples roaming the earth after the Crucifixion, or a Viking wandering south along the eastern shores of North America, have been based on the supposed native memory of a white-skinned, blond man. They remain hopelessly speculative. The historical King Quetzalcoatl is associated first with the Toltecs, an expansive, militaristic, and highly artistic tribe, which dominated Meso-America around AD 1000 from their capital at Tula, north of Mexico City. This was approximately coincidental with the collapse of Classical Mayan civilization, but still several hundred years before the rise of the Aztecs. In the skeleton story pulled from the ruins of the Chronicles, Quetzalcoatl is portrayed as a benign, compassionate, Buddha-like figure, who strenuously opposed human sacrifice, and could scarcely bear to harm any living thing. He was not, however, without his human enemies, who conspired against him. He was caught up in some scandal and forced to abdicate his throne. Quetzalcoatl fled to Cholula, then set off in a boat, promising to return some day and restore his dynasty. Several decades later, he died in the Yucatán peninsula, after having built his own funeral pyre. So much is known. All else is for the mythologist to re-create.

In the surviving artifacts, Quetzalcoatl is variously portrayed as the Morning Star, the wind god, the road-sweeper. But mainly he is a young man in a distinctive conical hat, sitting naked in heaven – presumably, the symbol of man's original condition and potential. In the hieroglyphic document known as the Vienna Codex, this alert youth sits at the feet of the Old Ones, the male and female aspects of the god-above-all. They give him four temples: one dedicated to the Morning Star, his personal symbol; one to the art of healing; one to the temple of the moon, his mother; and a last temple dedicated to the knot of Xipe, an obscure Toltec symbol. From this heavenly plane, Quetzalcoatl drops to earth via a ladder knotted like a sacrificial sash. This descent becomes crucial in most versions of the Quetzalcoatl myth. In the Legend of the Suns, for example, Quetzalcoatl is coaxed into a drunken orgy, during which he has intercourse with his sister. Afterwards, he repents carnal crime, descends into a stone casket for eight days, then builds his own funeral pyre and rises to heaven as the Morning Star. 'We have here,' wrote Irene Nicholson, 'on the one hand a physical description of the passing of the planet Venus below the horizon and its reappearance; and on the other a symbolic representation of a stage in the soul's pilgrimage.' This pilgrimage is also present in another version of Quetzalcoatl's descent, where his father the Sun is murdered by his four hundred brothers (the stars of the Milky Way). Quetzalcoatl journeys into the underworld to recover his father's bones and is thus placed in the middle of eternal struggles between the sun and the stars, day and night, light and dark. In still another version, the youth Quetzalcoatl steals the first tiny, precious grain of maize on his trip through the underworld. Upon his reascent to earth, maize becomes the symbol of his having achieved full manhood.

The casual reader of Mayan mythology quickly becomes dizzied by the complex, interlocking system of symbols, dates, and characters. It doesn't help that Mayan hieroglyphics have been reproduced, described, annotated, and analysed – but never literally translated. Yet in all the diverse versions of the Quetzalcoatl legends, two themes are comprehensible to a degree. First, Quetzalcoatl was not considered a full-fledged god, but rather a figure in transition, a man in the process of becoming divine. He is the human soul taking wing to heaven like the bird, and he is matter descending to earth as the crawling snake. He is virtue rising, and the blind force of aggression pulling man down. It was characteristic of Mayan symbolism to gather such opposites into a single figure, representing, for instance, matter and spirit, stillness and movement, the sacred and the profane. Where Christian civilization tended to believe such aspects were polarized and inimical, the Mayans saw not a synthesis of opposites, but a continuum. Man does not stand at the centre of the universe, but somewhere within the great universal hierarchy. Secondly, the theme of Quetzalcoatl's movement – from heaven to earth, from earth to the underworld and back, or from the earth back up to heaven after his death. Like a star, he's always in motion. But he becomes sovereign only by following his own cycle, obeying his own law of motion, not the fixed and determined laws of nature. By making himself move, he starts all creation moving. In a universe the Mayans conceived of as hierarchical and moving in fixed, eternally repeating cycles, Quetzalcoatl's travelogue must have seemed pretty exciting stuff: you are where you go. In this we hear the ancient echo of that great human migration which brought Asiatics into the American hemisphere at the end of the Ice Ages.

In whatever form the Quetzalcoatl tale was told, it was never simplistic, animistic, or, above all, primitive. It was not a neolithic, agrarian society's pictorial almanac of celestial movements any more than the Old Testament was a sheep-herding manual for the ancient Hebrews. Like all sophisticated myths, Quetzalcoatl operated on multiple levels. To the Mayan peasant farmer, it's true, Quetzalcoatl's course may have described the relation of the Morning Star to the sun, by which the farmer would know when to burn his *milpa*, sow his maize, and reap his harvest. To the Mayan nobility, the plumed serpent may have defined man's place in creation – and thus the nobility's powerful place in the hierarchy of human society, which was intended to reflect universal order. But to the Mayan priests and scholars, who learned to write, studied astronomy, kept the holy books, maintained a language of their own, and used legends to interpret the cosmos to their people, Quetzalcoatl appears to have been a young man probing the eternal puzzle of life on his journey to revelation: why are we here, in this particular place, at this particular time? As a universal image, Quetzalcoatl must be accepted at the level of faith. So were the *pochtecas* instructed in the codes of Chilam Balam to just follow destiny in Quetzalcoatl's footsteps, without explicit commandments, or even parables:

> You are to wander
> Entering and departing
> From strange villages . . .
> Perhaps you will achieve nothing anywhere
> It may be that your merchandise
> And your items of trade
> Find no favour in any place . . .
> Do not turn back, keep a firm step . . .
> Something you will achieve
> Something the Lord of the Universe will assign you.

Sunday rolled in on intermittent showers, but you can't let the weather skunk you on a thirty-day visa. Strategically, watching Quetzals in the tropical cloud forest could not be so different from observing any other bird: one part knowledge, one part patience, and three parts willingness to get wet. We knew by now there were Quetzals active in the Purulhá vicinity, and had a fairly good idea of their feeding pattern – it was a safe assumption that the rains did not alter the Quetzals' daily schedule. So we adopted a wait-'em-out routine of spending at least two hours in the field at dawn and dusk every day, birding from a stationary position. That way we'd have all day and all night to dry off at the Volunteer's place.

Next we chose the most potentially productive territory we'd encountered, an area of half a mile or so along the Cobán Road, including the grove of *aguacatillo* trees planted specifically to attract Quetzals. About half this territory was within the Biotopo reserve, the other half on the adjacent undeveloped properties to the north and east. It was a pretty spot. The Cobán Road cut a break in the canopy, allowing a clear view of forested mountain slopes above and below, where a startling variety of young trees grew thickly in the light gap. On the eastern side of the road, within the Biotopo, was a gushing falls, where the clear mountain water exploding on to the rocks sent up dense, sparkling sprays of tinted vapours. The first break in the weather came at 7 a.m. The fog slipped over the east face of the mountain ridge, and the sky opened briefly. The sunshine bore down like a laser beam, as birds rallied in the hot spots to feed. The first ones there were Azure-hooded Jays, a Central American cousin of our own familiar Bluejay, except dark blue-black in colour, with a light blue patch at the rear of the head, outlined in white. All jays share much the same personality. The Azure-hoodeds were screaming their heads off and acting very macho. They specialized in stealing food from other birds in raucous attacks. A flock of twenty to thirty Grey Silky Flycatchers started hunting in the tree-tops. We'd heard them singing their pretty but repetitive two-note song the previous evening, and it was a treat now to watch them dart out after insects and return to perch. Trim and sleek birds with a cardinal-like topcrest, their yellow-and-white flanks flashed with each swift loop into the air. Working my field-glasses through the trees, I began to pick up some of the better camouflaged, harder-to-see birds. There were two tanagers, one a Rufous-winged Tanager, the other a Blue-crowned Euphonia, both highly energetic birds exploring the trees for berries and seeds, and both feathered in lemon-lime with exquisite pastel patches, the very essence of brilliant tropical colours. The Euphonia's Latin name is *Euphonia elegantissima*, which might translate as 'Lovely-sounding and very elegant' – a nice name. Seeing its crazy-quilt plumage, I could only regret not being here when the Euphonias were singing. The most curious bird of the mixed flock was an Emerald Toucanet, wrestling with some small fruit about thirty feet above ground level. The Emerald is one of the smaller members of the toucan family, well known in American zoos for their clownish curved bills and acrobatic eating methods. As Skutch once observed, 'A long bill helps a bird reach food, [but] it creates a problem when it comes to swallowing.' The Toucanet's bill looked like a yellow-and-black-striped party nose tied on to its face. It swallowed by throwing its food into the air from the front end of its bill, tossing its head back, and catching the morsel in the deep part of its throat. Sometimes it missed. Or, by holding the food in its foot, swinging its head under, and tossing the food down its own gullet. It was full of outlandish ways to get around that Gogolian schnoz.

Then a trogon shot into view, crossing east to west from the downhill side of the road, and perched nicely in a tree on the Biotopo side. It was a Quetzal hen. She was plump as a dove, and rousing out her wet feathers, which made her appear plumper still. She had a rather demure greyish brown breast, with only a small patch of carmine red showing on her belly. But the white-and-black ribbing under her tail coverts was distinct and bright. I say a Quetzal hen, though there was an outside chance this was a fledgling, not an adult female at all. But the bird's calm and stately bearing seemed to me altogether mature. This was a very promising sign, because if she was indeed a hen of breeding age, there was a good possibility her mate and fledglings would be in the area as well. She perched for a full minute or so, blending so well into the sun-dappled foliage that her green back was all but lost in the leaves. Perhaps she was watching the bird-watchers. No others appeared. Soon she slipped off the branch, and with a quiet flutter of wings, disappeared into the forest on the upside of the ridge. In another five minutes the rain began again, and the only birds left in evidence were the *guardabarrancas*, spitting notes into the fog with the voices of long-lost, long-winded prophets crying in the proverbial wilderness.

Drenched to the marrow after an hour more of waiting, we decided to celebrate our first good view of a Quetzal with breakfast at the *comedor* called San Antonio, a kitchen attached to a gas station on the Cobán Road at the turning for Purulhá. The place had real adobe walls washed a slimy post-office green and real tables. There were even little bottles of toothpicks on the tables, and, wonder of wonders, a menu! But the menu turned out to be a figment of someone's imagination. The young waitress shuffled listlessly out of the kitchen after a few days. She wore a brown dress and flip-flops, and listened in morose silence as we ordered fried eggs, beans, toast, chillies, clotted cream, and coffee. Then she brought us *tortillas* and bowls of watery gruel: a few cabbage leaves and a gristly bone floating in lukewarm scum. The fleas paddled out to the bone in the middle of the bowl, and stood there wondering where to go, like the survivors of a shipwreck. As I took my first sip, I heard my buddy Mickey groan, 'How can you eat that?'

I couldn't. I called the waitress over. There was no use in being overly polite. 'Waitress, there are fleas in the soup,' I told her. I guess she had never heard that joke, because she didn't smile. She ignored me, and went over to a pair of truckers at another table. They cursed and spat on the floor when she presented them with the bill for their meal. One of them handed her a half-quetzal note. She went into the kitchen, only to return saying there was no change, the bill was too large. This is a common occurrence all over rural Central America, where many people live outside the money economy. Maybe she was hoping they'd say, 'Keep the change,' but no such luck. I saw my chance.

'I'll change it,' I said quickly, 'if you'll take this soup away and bring us something else to eat.'

'*No hay nada!*' the waitress snapped angrily: 'There isn't anything else.' For a moment our eyes met. She was only in her twenties. In her bloom she probably would have been considered a pretty girl, with black eyes and black hair setting off a light olive complexion. She seemed to be of mixed race – not dark enough to be a full-blooded Indian, not light enough to be a Ladino – but it was hard to tell. In any case, a bloom fades quickly in the dust and mud at the side of the Cobán Road, and instead of the hostility I expected to see in her face, there was a desperate expression that seemed to plead, 'Take me away from here,' though she said nothing.

We ordered Fanta orange sodas.

Meanwhile, a beat-up orange Toyota sedan had drawn up to the gas pumps outside. The car was packed like a telephone booth in a stuffing contest. One by one, eight fair-skinned young Ladinos uncurled themselves and stepped out. The girls were dolled up in long skirts and high heels, and wore heavy rouge, mascara, and lipstick. One of the guys was tubby, wearing a baby-blue joggers' outfit. They could not have looked more inappropriate, more out of place. The driver got out last, in a military jacket and dark aviator glasses. As he swung his legs around, he reached under the seat and brought out a sawn-off shotgun. They stretched and sauntered into the *comedor*, the driver bringing up the rear with his weapon raised, checking up and down the road. Inside, he laid that shotgun down on the table like a place setting and ordered the waitress to go put gas in the car. She didn't give him any tragic looks, I noticed. She did as she was told. While she was out at the pump, the gang helped themselves to sodas and bags of plantain chips. I noticed they never got a bill, either.

When the waitress came back inside, she let the screen door bang behind her. Everybody in the place jumped out of their chairs, watching to see whether the guy would go for his gun. Only a madwoman could have done that, and there she stood staring straight at me, as though it were my fault. Birds I can sometimes understand, Latin women, never. We put all the change in our pockets on the table, and made slowly for the door. The waitress was slouching against the back wall. As we went out, I heard her hissing at our backs, '*Gringos . . . gringos . . .* sons of whores!'

12

Later that morning we walked around Purulhá to see what Sunday was like in a small market town in the highlands: as in the days of the *pochtecas* commerce and religion are still intermingled; the sabbath was also market day. The Indians streamed down from hamlets in the outlying mountains, the women in their fanciest embroidered *huipiles* (blouses) and lovely striped shawls, in whose patterns can be read the precise town, tribe, clan, and sometimes cult the women belong to. Inasmuch as the Mayans did not live in such towns before the Spanish Conquest, it's thought that these local designs were adopted in the post-Conquest period, encouraged by the Spanish authorities as a system of immediate identification of the natives. But the patterns got so magnificently detailed that now only the Indians and a few anthropologists can read them. The men, on the other hand, were in their drab peasant trousers and shirts, as though intent on appearing nondescript. This has been the pattern in Central America at least since 1932, when El Salvador's General Maximiliano Hernández Martínez ordered his troops to shoot any man found in Indian costume, to burn all Indian settlements, and to destroy any artefacts of Indian culture. At least 30,000 people lost their lives in the space of a few weeks because they were wearing the wrong clothes. The event became known simply as *La Matanza*, the massacre. Fifty years later, you don't see men in El Salvador wearing native dress. Fewer and fewer in Guatemala risk it. Why make yourself a target?

Whole families walked the roads into Purulhá without shoes on their feet, the children scrubbed clean and their brilliant straight black hair washed and neatly combed. The smaller ones rode their parents' hips and shoulders. The older ones trailed in single file behind, helping to carry the produce their parents were bringing to market. They were remarkably well-behaved children, never breaking ranks to throw a stone, or skip ahead, or climb a tree. You never saw a parent scolding a child, or, perish the thought, striking a child. Misbehaviour and punishment did not seem to be part of the Indian vocabulary. They walked along in a quiet, dignified rhythm, as though lost in a religious trance.

Two foot soldiers of the National Police trudged the back road into town, drooping under the weight of their old M-1s. They stationed themselves on the veranda of the municipal building, where they could keep an eye on the market shed and still see across the barren plaza to the church. The town boys brought them Cokes in plastic Baggies. The soldiers looked very officious, and very bored, but not especially threatening. They did not budge from their chairs on the veranda. Across the way, the heavy worm-eaten doors of the old town church were wide open, but no one went inside except for mongrels and stray pigs, which lay down on the church floor to keep cool.

'Where's the priest?' I asked a woman.

'*Saber*,' she said.

'Where's the priest?' I asked another.

'He's dead,' she whispered, crossed herself, and hurried away.

'The priest – where is he?' I asked yet a third.

'He fled.'

'He didn't flee,' interrupted a fourth. 'He's here, but he's hiding.'

No one knew for sure what had happened to the priest, only that he wasn't saying Mass any more.

In the market-place, posters had gone up, advising the citizens: 'Guatemalan: respect for authority is part of your tradition'. There was very little for sale in the market, only some potatoes, cabbages, bananas, onions, chillies, blocks of unrefined brown sugar, green herbs, and the cigars Indian girls smoke to bring on visions of their future husbands. There was nothing to do but drink Fantas and watch the authorities watching the Indians, so we walked to the edge of town and hung out for a while with the civil-defence patrol. They were busy checking the papers of everyone who came into and went out of town, photo IDs that every Indian over sixteen is obliged to carry at all times. The members of the civil-defence squad were volunteers on a mandatory basis. Every able-bodied man and undrafted teenage boy was expected to go on duty one day per month. It wasn't healthy to refuse. The members said the patrols had been instituted in October, after a band of guerrillas had slipped down from the mountains and raided a Texaco station a few kilometres up the Cobán road. There was talk of rebellion then in Purulhá, but the army had arrived to 'pacify' the area. The situation was calmer now, but just how it got calmer was not very clear. Several priests suspected of sympathizing with the rebels because they preached the theology of liberation had disappeared from towns in the vicinity, but whether assassinated, run out, or frightened away, none of the patrol members could say. They spoke of rumours that an *aldea*, or hamlet, in the hills had been chosen by the military to make an example of, the buildings razed, the population exterminated for supposedly aiding the rebels. But none knew the *aldea*'s name, or its exact location: the civil war in Guatemala is made up of just such 'battles', which the military wants to keep secret and the Indians would rather forget. A war you can smell, and sometimes hear, but as an outsider almost never actually see in progress. Or to put it another way, a war easily hidden. All the military needed was a roadblock, a gasoline bomb, and a bulldozer, which is perhaps why a conflict as bloody and violent as anything taking place in El Salvador or Lebanon goes on in rural Guatemala with so few from the international media paying attention:

During an interview carried out in April by a foreign journalist . . . civil-defence patrol members in Baja Verapaz admitted that they had been involved in such atrocities. They stated that they acted under the orders of the military commanders who instructed them to consider as 'involved' anyone they found over the age of twelve in areas or houses considered suspicious by the commanders. They were told to seize such people and kill them. Even younger children, if they too were felt to be 'involved', were to be summarily executed. The testimony stated that until recently, the women had been left alone in the houses when the men were taken off, but that now the women were being routinely raped, even those that were pregnant. One member of the squad told his interviewer that in one case a woman was raped five days after giving birth, when she had left her home to bathe the baby. He also reported having seen people drowned and mutilated, and said he'd seen several people's ears being cut off . . . The soldiers who directed these civilian squads were, according to the informant, also young Indians, obliged by their commanders to order the civilian defence squads to commit such atrocities. Another member of the patrol told of seeing a man who tried to escape being recaptured. All his muscles were cut and gunpowder placed in his navel and set on fire. The victim's eye was put out, and his skin then peeled off. The soldiers joked that they were going to have a barbecue.

This from an Amnesty International Report: the legacy of Alvarado endures.

It was a few months before that the army had gathered the towns-people together in Purulhá, and told them to form a civil-defence squad to protect their town. They hadn't armed the civil guard, but had collected all privately owned rifles in town, then given them to the guard for their patrols. It was a fundamental element of the general's counter-insurgency programme known as 'Beans and Bullets': beans if you behave, bullets if you don't. Everyone was be-having, but still waiting for their beans. The Guardia had effectively sealed off the town from guerrillas, strangers, and neighbours alike, isolating the population so that the military could withdraw its troops to the better-defended provincial capital at Salamá, about forty kilo-metres away. The civil-defence patrols cordoned the town more effectively than uniformed soldiers: the soldier of an occupying army can never be your friend. The men of the civil-defence squad, on the contrary, were literally friends, neighbours, uncles, sons. What could be more diabolically economical than to make the chickens guard the chicken coop? To join, feed, or even discuss the rebels now meant a complete break with one's entire life – village, *milpa*, family, and church, all the things that have helped the highlands Indians survive the centuries of travail that began with the Conquest. It probably meant a one-way flight to the Mexican border, and it also meant that civilians caught up in counter-insurgency operations or guerrilla raids would have a dangerous time fleeing. The civil-defence squads were supposed to report every strange face. Just to run away from the violence was now to be considered 'involved' in it.

It was working. Cut off from the population, the rebel groups operating in the highlands were taking a licking and descending into banditry. Whatever raids they were able to carry out now concentrated on easy targets, the quick destruction of property, procuring supplies. 'They wanted food, and medicines, too,' said the kid at the chicken shack on the Cobán Road where the raid had taken place. Next door was the remains of the Texaco station the guerrillas had burned. He said, 'Two of them were nurses.'

'Were they local people?' I asked.

'No, they were foreigners. North Americans.'

'What? No, they couldn't have been North Americans. *We're* North Americans.'

'They were Nicaraguans and Cubans, then,' he corrected himself dubiously. 'I heard them with my own ears and recognized their language.'

'Were you frightened?'

'No. I was valiant.'

'Did you give them what they wanted?'

'Don't have medicines. I told them. They just took food, then.'

'Chicken?'

'Yes. They ordered fried chicken to go.'

The chicken shack was decorated with Mayan heads painted in lurid, psychedelic acrylics on the grubby green walls. There was a pool table on the patio, and a loud, fuzzy jukebox playing inside. There were a million greenbottle flies around the food. The guerrillas had burned down the wrong place. Having told us the story of his heroism, the kid wanted to know if we were *turistas*. 'I have always wanted to meet some tourists here on the Cobán Road,' he declared.

I told him we were looking for Quetzals. Had he seen any?

'Sure. In my pocket.' The inevitable joke. We didn't laugh. He watched us for a moment suspiciously. Then he leaned over the chicken bones and murmured, 'You know, I don't think we'll be seeing any more Quetzals around here. The Quetzal is the bird of freedom. But here there is no freedom.'

13

A hard rain slashed through the night, tumbling the pebbles and mud of Purulhá down towards the swollen stream spinning wildly round the edge of town. The storm tapped a tinny calypso beat on the corrugated roof of the Volunteer's house, but there were intervals of lighter showers, and interludes when, although no drops fell directly overhead, you could hear the rain pelting the roof in the next yard, or charging up the valley like horse cavalry: half a mile away, coming closer and closer. Each time the Byzantine twists of the downpour swept back our way, you could smell it coming, whipping the chill musk of the cloud forest ahead. Then you lay there waterlogged on the concrete floor, listening to the oldest sound in American civilization: fat drops slapping against the broad corn leaves, hissing in the banana trees.

At 5 a.m. the night finally calmed, the clouds cleared away, and dense stars packed the saturated sky. The Milky Way cut a bright swath through the heart of the heavens. 'Time to go to work,' said Mickey. We jumped into our soggy jeans, munched bananas, and pointed the Toyota, full of piss and vinegar even at this hour, towards the Biotopo. In Purulhá, the people were already on their way to work, too. The rushing rut of muddy water that passed through the centre of town was strung with the ghostly figures of Indians. The gate at the head of town had been abandoned and left open – not even the civil-defence squad was intimidated enough to stand out all night under the thunder and lightning. It was as if the weather itself had accomplished a modest reintroduction of sanity to Guatemala.

In the chill pitch of the Cobán Road, shadows pedalled bicycles up and down the mountain passes, or humped along on foot, bent under crushing loads. The daily pre-dawn bus for Guatemala City broke wind as it slowed down beside the stop, the scrollwork insignia along its side illuminated by the running lights: Monja Blanca, the White Nun, it said. It was a funky old streamlined Greyhound, which must have migrated south of the border in the fifties. The baskets and cardboard boxes piled high on top of the bus resembled a moving skyline in miniature of some great Mayan ceremonial city, flat-topped pyramids and arched temples hulking against the indigo horizon.

Nor had Fernando and Francisco waited out the storm at the Biotopo. They'd gone home to sleep in Purulhá. The gate of the Biotopo was shut and tethered. As we waited there, the night receded against the damp gloom of dawn. Blazing Venus descended slowly towards the navy flanks of the Verapaz mountains. Then the *guardabarrancas* opened with their skirling cascades of notes. The trail crew was arriving.

'*Buenos días.*' ('Good morning.')

'*Que le vaya bien.*' ('That it may go well with you.')

Soon enough, Fernando pumped up the road on his aged black wheel. Francisco was mounted behind, helmeted in his father's yellow hard hat, which beetled down over his eyes. The 'Resource Guard', as

Fernando's job title describes him, greeted us with a silent wave and proceeded towards the gate. But as they passed, the impish little Francisco flashed a wide, sly grin – his arm shot out, pointing back down the road in the direction they'd come from.

We needed no further instructions.

Grabbing glasses and cameras, we dashed down the road in the gathering greyness. A quarter mile away, we met an elderly *campesino*, trudging barefoot with his heavy-headed hoe over his shoulder.

'Certainly I saw them,' he replied to my question. 'Three Quetzals together, in the trees right next to that little cabin over there.'

'When?'

'*Ahorita!*' he replied, the diminutive form of *ahora*, or now, which is the way Central Americans describe all the variations of time on the margins of the more precise, but less flexible, English term.

The road by the unoccupied cabin was within the territory we had under surveillance. The valley falling away below the road was still swept by rolling mist. We took up observation posts a few hundred yards apart on either side of the cabin and crouched down silently, trying to remain alert despite the lack of morning coffee. Something red and white flashed at the corner of my eye, and when I turned, two Quetzals fluttered across the road heading south towards Mickey's position. With their brilliant crimson bellies and snow-white patches under the tail coverts, they were as identifiable as traffic lights – a pair of hens, both lacking the male Quetzal's fancy tail plumes. I froze in place, fearing that the slightest movement, either lifting my binoculars or signalling Mickey with my arm, might scare them away. One bird entered the upside thickets, and I saw it no more. The second Quetzal, however, landed in a cecropia tree well within eyeshot, much as the one had on the previous day. It was impossible, of course, to identify this individual as the same Quetzal, but she shared the other's aristocratic bearing as she perched, gazing around alertly. I hadn't long to wait before she leaped from this perch, and with a graceful upward swoop, reached out for some small fruit. As she did so, she hovered in mid-air for several seconds, wings beating rapidly like an immense hummingbird while she fed. This was enormously thrilling to watch. The bird's red-and-white front flashed on and off superbly with each flap of the wings. Then the Quetzal dropped down to perch again, in the understated modesty of her plumage. When the Quetzal landed, I could now see that its back and wings were covered with feathers of an olive-green colour, not the brilliant and legendary green. This added to my impression that the bird was a hen. She repeated this vivid and robust feeding manoeuvre several times at several different trees, flashing her colours in rapid succession, like a signalman's semaphores. In the intervals between these swoops, the Quetzal rested without twitching a muscle or ruffling a feather, as though these bursts of activity were to be balanced with equal periods of absolute stillness. I thought then of the Mayan gathering of opposites: intense display of colour and near invisibility, pure movement and pure rest. After perhaps ten minutes of feeding, the Quetzal retreated uphill into the forest, where the first bird had flown, and in the same direction as the Quetzal we'd observed yesterday.

Breathless, and in a state that verged on the transcendent, I raced down the road to find my partner. 'Did you see them? Did you? Did you get a good look?'

'Slow down, buddy,' laughed Mickey. 'You'll have a heart attack.'

'I'd die a happy man. I had two back there. Looked like two females. The old man said he'd seen three birds.'

'Yeah, it's too bad,' he shook his head in mock sorrow. 'You should have seen the tail plumes on that big male I saw up there. They must have been three feet long. The thing passed about ten feet over my head. Guess you really missed out, huh?'

I held my tongue, but in a way I was glad the Quetzals I'd seen had been *hembras*. That way, I still had the sweet anticipation of seeing the gorgeous guy himself.

14

It didn't take long. Twenty-four hours later we were there again, same time, same place. The weather had taken a sharp turn for the dismal: a cold, steady rain that plodded on hour after hour like an unimaginative form of psychological torture. Francisco crouched on the bank across the road, watching the crazy gringos. He held a piece of a cardboard box over his head as a rain hat. The Volunteer was along, too, hefting a gargantuan pair of rubber-armoured military-surplus binoculars. An hour crept by. Then an hour and a half. There was nothing to do but hunker down and wait in the miserable downpour. When the sun finally managed to break through, the territory quickly became its usual mayhem of species. The birds poured into the light gap, chattering greetings to the sun. The Emerald Toucanet was back. So were the tanagers, the flycatchers, the jays – the whole gang, by now familiar feathers to us. A Quetzal hen crossed the road first, but she was swiftly followed by a large male, his tail plumes waving elegantly behind him. His colours ignited in the sun, emerald green changing to golden green, then to an iridescent green that defies any name but Quetzal green. He crossed west to east with an easy, undulating flight, tossing his tail plumes playfully in smooth arcs, before setting down to perch.

'The crotch of the cecropia,' I indicated to the others quietly.

'*Guarumo*,' corrected Francisco, crouching on the bank and taking it all in. *Guarumo*, then, was the local name for the cecropia. It is a common tropical tree, which grows as a pioneer colony wherever primary forest is eradicated, as here along the Cobán Road, thrusting up quickly to catch light, and only later putting down overground roots to anchor it. It forms a kind of hard, fruity spike, and the male Quetzal repeatedly bobbed, fluttered, and plucked at these, all with noble precision. I could not quite tell whether the bird was taking insects or seeds from the spikes. Its feeding behaviour, however, was very much like the female's, but with this difference: the male Quetzal left the branch backwards, taking off in reverse, as it were, probably to prevent its long tail feathers from sustaining damage as he rose. A wonderful and athletic manoeuvre. Much literature on the Quetzal reports that males often break, or severely damage, their plumes in the course of the nesting season as they go in and out of the nest – they may be called *machos*, but male Quetzals share nesting duties, sitting on the eggs and feeding the newborn, equally with their mates. But at least this Quetzal was making a conscientious effort to protect his plumes, the golden-green plumes that once adorned the heads of Mayan lords. His train was in exceptionally beautiful condition, especially considering that this was so soon after the wear and tear of nesting season.

A second male with an even more prodigious train then crossed the road directly overhead. Behind him came three more Q-birds. One was a hen. Judging from their smaller size, their slightly motley, buffy feathering, and also from the way the male in front and the female at the rear flew so solicitously close to them, the two in the middle were more than likely fledglings, pretty fresh from the nest. So now there were six Quetzals in view, apparently two sets of parents and the kindergarten of two. The adults brought the fledglings to the *aguacatillo* trees, where the young were mastering that same upward swoop, mid-air hover, and gobbling of fruit, the Quetzal's aerial signature. They did not stay there long, but were soon dancing back and forth over the Cobán Road. They looked as if they were having a smashing good time. I was practically unaware of time, frozen in ecstasy: I had never seen a more deliciously feathered creature. Then the prong of sunlight poking through the drab rain closed. And our Quetzals vanished into thick, new fog.

15

Later that morning I had an apology to make – a moment of monumental embarrassment to complement our modest ornithological triumph (the Mayan gathering of opposites again). I'd bought a machete for one quetzal in the market-place at Cobán, with the intention, mostly, of using it as a garden tool back home in New Jersey. Its long blade gleamed in a tantalizing way. The metal was stamped 'Made in El Salvador'. Now, Don Lázaro kept a fine grove of banana trees out behind the Volunteer's house, a profuse clump of suckers, some as high as thirty feet, which happened to surround the Volunteer's privy. The earth there must have been magnificently fertilized by the prodigious production of the Volunteer's distressed bowels. Bananas are called trees, but they are actually a herbaceous shrub. Instead of a woody trunk, their core consists merely of the enfolded leaves, which shoot up as they open out into those extremely long, gracefully arching leaves so characteristic of the American tropics. Within six months they have grown to full height, flowered, and set fruit in heavy, low-dangling bunches. Commonly, banana trees are cut down by machete along with the bunch of bananas the plant produces, as new suckers spring up to take its place. They are ubiquitous dooryard trees in the Central American highlands, producing immense amounts of fruit but requiring practically no care, and a family homestead would no more be without its banana grove than without its *milpa* of corn. Hummingbirds often build their nests under banana leaves like a tiny hanging basket, and the trees are frequently visited by Banana-quits, a dainty bird with handsome white eye stripes and a lemon breast.

I was on my way to the privy, slashing with the new machete through the dried brown banana debris on the ground, when the blade, to my foolish amazement, suddenly sniped a prime specimen and brought it crashing down. An enormous bunch of immature green cooking plantains whooshed past my head and hit the dirt right in front of the outhouse door, ruined. I had to climb the hill to Don Lázaro's house and confess my crime.

He wasn't home for dinner yet, but his wife took the loss (and my apology) in good humour. When she related what had happened to the old lady minding the kids on the porch, they both burst out in gales of laughter. Don Lázaro soon putted in on his Yamaha. He took the news in equally good form, shook his head, and said not to worry about the loss of the plantains. He wouldn't hear my offer of restitution. Still, I was too embarrassed to accept their kind invitation to stay for the noontime meal, which Señora Don Lázaro returned inside to finish preparing for her husband. Out on the porch, Don Lázaro made a gracious transition to another subject. '*Ayybuen*, I used to be a fair hand with a machete myself,' he said. 'I worked as the supervisor of trail construction at the Biotopo.'

I didn't waste any time inquiring about his work there. A number of such good jobs were created when the Biotopo was first established in 1978, he began. Local men were hired to cut walking trails through the cloud forest, and to reforest damaged areas (although mainly with pines, which are not much good to the Quetzals; their beaks aren't hard enough to crack pine-cones), as well as to construct the guardhouse, camping area, parking apron, and so forth. It was one of those happy occasions when a project of so-called development coincided not only with good conservation policy, but with the creation of productive, decent-paying jobs, too.

At the time of the Biotopo's inception, it was actually thought by many that the Quetzal was either already extinct in Guatemala, or on the very verge of extinction. The bird had disappeared entirely from its traditional range in Quezaltenango, Huehuetenango, and Quiché, as had the cloud forests *per se* there, under the relentless pressure of slash-and-burn farming, timbering, and livestock grazing. Nation-wide, game wardens were confiscating three or four Quetzal skins per year from poachers, which is not many. It was gossip in Costa Rica at this time that the Guatemalan military government had made discreet inquiries about buying some Costa Rican Quetzals for importation into Guatemala, in order to mask the scandal of the Guatemalan symbol of nationhood and liberty becoming non-extant.

The Biotopo had been the work of many, Don Lázaro said, but the man who conceived this system of biological reserves, who became the driving force in lobbying the Guatemalan government into sup-porting the idea, was a professor of chemical engineering at the Univer-sity of San Carlos in Guatemala City, Mario Dary Rivera. As the first superintendent of workers at the Biotopo, Don Lázaro worked directly under Professor Dary. He called him *Licenciado*, which literally means lawyer, but is used throughout Central America as a term of respect. 'Licenciado Dary was the person who felt most deeply for the Biotopo,' Don Lázaro said. 'He didn't live here full time, but he came every weekend and on vacations. Sometimes he brought his students to work by his side. Sometimes his whole family. Every week he would be telling me, "This week, build a trail here," or, "Now we'll do the resource plan." A very fine person, very intelligent, totally dedicated to conservation, not only of the Quetzal, but all the animals. I re-member, one time, he told me to find a person from town to hire as a guide, because he had some North American scientists coming to investigate the Biotopo territory. He was very specific. He said he didn't care if the person was young or old, man or woman, Indian or mulatto, just so long as the person understood not to kill any animals. So I hired a man. He came on the job, and one day the North

Americans came. The man went out with them into *la selva nublado*, the cloud forest. They were gone most of the day, but when they came down from the mountains, the man I had hired had killed a green snake, about four feet long, with his machete. Licenciado Dary was sitting in that little thatched visitors' centre, and when he came out and saw the serpent, he got furious. The man said to him, "But, licenciado, snakes are dangerous around here. I was only trying to protect the North Americans." Dary picked up the snake and inspected it. He said it was a non-poisonous kind that ate mostly vanilla beans. He told the man, "I can't have anyone working here who doesn't respect living things," and fired him on the spot. That's the way he was, Licenciado Dary. He wouldn't let anyone kill a single animal in the Biotopo. He was very strict about that.'

I asked, 'But why do you say that's the way Licenciado Dary "*was*", Don Lázaro? Doesn't he still come around from time to time to see how the Biotopo is functioning?'

Don Lázaro paused for a moment. Then he said, 'You mean no one told you?'

'Told me what?'

'They killed him,' he said. 'Gunned down in the streets of Guatemala a few years ago.'

16

Don Lázaro was right on the money when he described the Quetzals as 'rather meticulous' in their habits. We began to see this small flock of Quetzals regularly. Every morning just after dawn they moved through the terrain at the edge of the Cobán Road, their behaviour well routinized: 45 minutes of feeding on small fruits, 15 or 20 minutes of aerial play, and then to a branch for a rest, 10 minutes or so of calm preening before they moved off in unison. Only the size of the flock varied from day to day, between three and six. Moments of sun brought them out to display their most brilliant colours and stunts. The more usual rain slowed them down, calmed them. But the wet, cold weather did not disrupt their activities. They seemed entirely well adapted to the wet climate.

It was a charming family scene, as the adult Quetzals accompanied their young of the year on the latters' initial weeks away from their nests. They were bringing their offspring down to a slightly lower elevation than their nest sites, to a place just below the perpetual cloud line shrouding the mountain summits. I could not discover why they behaved so, only speculate. Perhaps it was because the young Quetzals, with their waterproof adult plumage not yet fully grown in, were more vulnerable to the weather at higher elevations. Maybe there was simply more food down below, where the rains brought on fruiting in so many trees along the roadside gap in the forest canopy. We assume that most birds have their reproduction cycles biologically fine-tuned in order to reproduce at the time of year when the most food is available for the nestlings. Why not also their feeding pattern? Then again, it was certain that the Quetzals covered a much larger feeding territory in the course of each day than we could keep under observation from our fixed point. Did they begin their schedule, perhaps, in the gap along the Cobán Road, and, in a sense, follow the sun as it rose higher in the sky? That would be very Mayan of them.

Whatever the case, it was clearly a happy period in the Quetzal's life cycle, and a fine one to observe. The parent birds were newly released from the fatiguing responsibility of constantly bringing food to the nest for their young. The young were testing their wings – learning how to be Resplendent Quetzals, under the watchful guardianship of their parents. Many keen students of avian behaviour believe that birds experience joy in the culmination of their biological tasks. A strong sense of parental pride, on the one hand, and a youthful thrill of newfound freedom for the immature Quetzals on the other, may have been the emotional basis of the most spectacular aerobatic displays the Quetzals put on each morning, dancing, dashing, sparkling, fluttering over the feeding grounds. Their movements were so energetic it was impossible not to attribute them to *joie de vivre*. I went back to the books on this point, and found that Alexander Skutch had noted something quite similar in the original fieldwork on Quetzals he accomplished in the 1940s.

From March to July the male occasionally flies up well above the tree-tops, circles in the air, then descends into the sheltering foliage. His flight on these sallies is strong, swift, and direct, often with little of the usual undulatory motion ... As he soars up in the air, he shouts loudly a phrase which at various times I wrote as 'wac-wac-wac-wac-wac-wac', but often as 'very good, very good, very good'. These spectacular ascents spring from pure exuberance; they appear not to be used in courtship and are certainly not for the purpose of finding food. I know of no other trogon, nor any bird of the dense tropical forest at whatever altitude, which indulges in similar exercises. The high flights of quetzals are another expression of the bounding vitality that has produced their elegant plumes, the approach to male coloration in females, although this is unusual among trogons, and the long breeding season, extending into the inclement months.

Back on our familiar, dank concrete pallet at the Volunteer's house, however, gaiety was as scarce a commodity as sleep. I flopped and shivered, trying to make some sense of these days of extremes and Quetzal sightings. Of the bits of information, the modest insights, the shards of myths and stories I was picking up. Of the enervating clash between a tropical landscape full of interest and bounty, and a human condition full of tensions, fears, privation, degradation, and death. I wrote in my journal:

What did Skutch mean exactly by 'bounding vitality'? How could vitality produce elegant plumes? Or a female almost as admirable in beauty as the male? Or a long breeding season? His concept borders on the metaphysical – that each creature contains within itself a certain amount of life force, or elemental natural energy, which is directed by the evolutionary process into various facets of existence: sensuality, brilliant colours, physical vitality, and the heightened emotional rhythms of family life we have witnessed. Does the Resplendent Quetzal, for reasons never to be understood, have *more* of this life force than other animals? Can this be why ancients and moderns alike have been so enchanted by this bird's extraordinary driving elegance? Why has the Quetzal managed to captivate the soul of tropical America through so many centuries? I'm attracted to this notion. It reunites matter and spirit, natural science and tropical cosmology, in a very Central American way. Such a natural-spiritual power rises above the griefs humans have dealt each other in these hills. I have an idea its force can survive the species itself. Even now, in the waning days of the Quetzal's existence on earth, new legends of the bird are generated to an astonishing degree. We've heard two good ones in the past few days. Everyone in Guatemala, apparently, recalls how, just last year, on the very day the pope arrived for his historic visit to this country, a fully plumed male Quetzal unaccountably flew through the window of a humble woman's shanty on the outskirts of Guatemala City. That's a lot of mileage from the nearest cloud forest. The bird shared the front page headlines with His Holiness. They were considered equally important religious events – maybe related events. The poor woman didn't have any idea what to do with the Quetzal flying around her shack. She had enough sense to call the zoo. Two students of the late Mario Dary came on the case. They collected the bird in a cage, and kept it temporarily in a room at the zoo. The Quetzal thrashed against the wall trying to regain its freedom, and wouldn't touch the food they tried to give it. The woman took this as a sign that she should bring the Quetzal to Quezaltenango to deliver it to the pope. However, inasmuch as the Quezaltenango region has been deforested, and the publicity would probably kill it, the biologists drove the Quetzal up to the Biotopo and released it. I can only wonder how that story will be told three thousand years from now.

Here in the highlands, they still tell of the memorial service held for Mario Dary on the first anniversary of his assassination. More than two hundred of his friends, colleagues, and students came to pay homage. Don Lázaro was there in the crowd, and swears this took place. So was Fernando. He says he saw it, too. Just as the assembly began to sing the national anthem in Dary's honour, a glittering Quetzal emerged from the cloud forest and flew to the top of a tall ceiba tree overhead, where it perched in all its great dignity and beauty, waving its emerald plumes in solemn salute. It sounds to me like the sanctification of a martyr.

17

Soon the road into town washed out. Twin shifts of Guardia were assigned to stand in the rain at the rear gate. There wasn't even a tree for them to hide under. Days and nights together the wind chucked down water in gelid sheets, and the fingers of lightning sizzled across the mountains, and the thunder boomed. The small morning sounds of Purulhá had to wriggle and fight their way up out of the great whipcord of the storm: the staccato barking of dogs, the thwack of machetes cutting kindling, the squawling of cold, wet babies.

There wasn't going to be any 'little summer' this July. Our field operations were suspended. In a few days we were scheduled to meet Alfredo Schlehauf at his coffee plantation Finca Remedios, and we needed a brief respite to dry out. And to put something in our stomachs beside *tortillas*, bananas, and Fanta orange sodas. In addition, I wanted to find out more about the Biotopo's founder, Mario Dary. That could only be done in Guatemala City, as our friends in the highlands did not know about his untimely death, or wouldn't talk about it. There was just time for a quick trip to the capital, so one morning we split our last banana for breakfast, thanked the Volunteer, presented him with all our toilet paper in a short ceremony (no Quetzals appeared), and loaded up the Toyota four-wheel, standing cleaner than new after a non-stop seventy-two-hour washing. An hour after dawn the rain eased a little, and we left. Our last view of Purulhá was the civil-defence squad at the roadblock on the back road out of town. It was Fernando's turn to stand sentry. He looked rather ashamed and bedraggled with the rusty single-shot rifle over his shoulder, and the water dripping from the brim of his old straw hat.

'Eh, Fernando, will the sky clear today?'

'Let's hope so,' he said. By his side, as always, sat the grinning Francisco, unconcerned by time or tide – or by the wet ground he was sitting on, for that matter. He was studiously scratching in the small notebook we'd given him as a gift, with instructions to keep count of all the birds he saw and to let us know about them when he became an ornithologist.

'So, Francisco, how many Quetzals?'

'Twenty!'

'And how many Zopilotes?'

Francisco giggled, and nooked his head into his father's knee. 'Twenty million!'

As soon as we got back on the road, we re-entered vulture country. We drove till we got to the provincial capital of Salamá, the town where the Nobel Laureate Miguel Ángel Asturias spent his childhood in the early years of this century, sitting on the knee of the family maid Lola Reyes, listening to her magical Indian stories of Quetzals, jaguars, and bats that talked. I couldn't believe this was the sheltered provincial town that had given the world one of Latin America's foremost writers – *Gran Lengua* (big tongue, hence great interpreter) of Mayan mythology. A carpet of red dust rolled out to meet the gritty little Toyota, and the road was lined with tacky mom-and-pop bars. The whores filled the open doorways, trying to lure in whatever custom happened to pass. There couldn't have been much. They smacked their lips, rubbed their thighs, rotated their pelvises, jiggled their breasts, squeezed their nipples, scratched between their legs, and shouted out come-ons – 'Special price for you, gringo' . . . 'real cheap here.' The Zopilotes watched inertly from the corrugated metal roofs. Reality stopped at the edge of town, and stood there swaying in the liquid heat like a boozed-up gunslinger.

By the time we turned up towards the plaza, a guy on a small Spanish horse was galloping ahead of the car to herald the advent of two North Americans. Men in cowboy hats, cowboy shirts, and cowboy boots came out to the high kerb of the sidewalk and slouched picturesquely against their pick-up trucks to watch our one-car procession. Only a few actually had six-shooters strapped to their hips. The corner music store put on a cassette in our honour and wrenched it up to full volume. You could hear it all over town. It was a tape of old Western TV show themes: 'Bonanza', 'Gunsmoke', 'The Marlboro Man', and 'Maverick':

> Who is the tall, dark stranger there
> Maverick is his name
> Riding the trail to who knows where
> Luck is his companion
> Gambling is his game

The army garrison overlooked the plaza, as usual. It was a new masonry building with heavy sandbag reinforcements out front. Some 'subversives' had run a car loaded with explosives against the old garrison, demolishing it. The soldiers stood around out front, keeping an eye on traffic, but there wasn't much else to keep an eye on: chickens pecking the gravel in the dooryard of the closed church, men throwing dice out beside the soda stand. Mickey went over to the soldiers, but when he lifted his Nikon to shoot, they panicked. They rushed forward, menacing with their automatics, and pitched him backwards with their hands. I held my breath until he returned across the street. It was the first time I'd ever seen my partner lose his cool.

'You see that white guy in the camouflage beret?' he ranted.

I looked back across at the garrison. A stout, red-faced, middle-aged man in baggy camouflage fatigues was just stepping back into the edifice out of view.

'He's the one who gave the order,' said Mickey. 'He's an Israeli adviser. I heard him speaking English.'

'Guess he didn't want his picture taken.'

'I guess not.'

'Look, forget it,' I said. 'Come on. Let's walk in the plaza. You can get your shoes shined.'

This seemed to pacify him. 'OK. Remember, walk tall, buddy,' he said, and marched off, dragging one leg like a damned British colonel.

The old-fashioned plaza sighed exhausted charm. It was divided into prim quadrants and still boasted a few dilapidated palms with whitewashed trunks. Their frowsy fronds had been taken over by flapping vultures and chattering grackles, which had no trouble chasing the larger, but far less aggressive, vultures away from their roosts. The ground stunk of their white droppings. Small islands of hibiscus and lilies floated in bare, worn dirt patches, and the carcass of an old Spanish cannon pointed the way to the boot and tack shops across the street. Some were closed for siesta, some were just closed. Where the concrete paths met in the centre of the plaza stood a peeling green concrete structure about twenty feet tall, maybe an ex-fountain, an ex-bandstand, an ex-monument, or something. The Indian shoeshine boys were coiled in a serpentine heap on the steps of this green thing. They barely stirred at our approach. One boy finally raised his head and rubbed his eyes. He wasn't rubbing the sleep away, but perhaps trying to rub the shabby town away so that he could go on dreaming.

'Shine, *señor*?'

Mickey put his running shoe up on the box, and the boy got right to work smearing thick black liquid on the white nylon.

'What is that thing?' I asked the boy, pointing to the green concrete.

'Eh,' he said without looking up. 'That is what we call here the Tortilla Smasher.'

It was just then that the Apparition appeared. He was as young, blond, clean-shaven, close-cropped, sanitized, and square a gringo as you'd come across at your front door in the States on Sunday morning, pushing his version of piety. He was dressed in an absurdly clean white, short-sleeved business shirt, pressed blue slacks, and a thin black knit tie. His narrow lips drained of blood to the colour of overdone liver as he said, 'When I saw you two get out of the car, I just knew you were from back East. How are things *back East*?'

The phrase tumbled awkwardly across the plaza. It was some kind of code word, but I didn't get it.

'Well, the Pony Express went belly up a few months back,' Mickey mocked him openly.

The Apparition gave off a self-conscious whiff of disconcerted innocence. 'Sure is good to meet someone from *back East*. What brings you to these parts?'

'Just passing through,' I gave the obligatory reply.

'Really? We don't see too many Americans here. Don't tell me – you guys are reporters, right?'

'No.'

'Look, I'm glad I ran into you two. A lot of Americans have the wrong impression of this country. You hear so many false lies about Central America.'

'As opposed to true lies?' I ventured.

'Well, like for instance, all the stories about killings down here. I haven't seen a single thing like that. This country's really at peace now. The general has done a heckuva lot for these people, but it never gets reported in the States.'

'What are you doing here?'

The Apparition said he was a missionary for a Protestant evangelical sect and had been in the highlands a few months trying to convert the locals. It was slow going, but it wasn't going too badly. 'I'm telling you, these people are really ready to commit themselves to Jesus. We've had some really great meetings. The general's a Christian, you know. You just can't believe all the things you hear. Why, if there weren't freedom of religion down here, our mission wouldn't be able to come here and hold meetings in the first place.'

'We've been noticing how many of the churches are closed. Seems like some priests have disappeared.'

'Who told you that? I haven't heard anything about that.'

'Well, you certainly have your work cut out for you.' I tried to end the conversation.

The Apparition wouldn't go away. 'Oh, it took some getting used to here. I'm from *back East* myself.'

'Where's that?'

'I've been all over. Some time here, some time there. Wherever Jesus needs me, that's where I go!'

'And where did Jesus need you before Guatemala?'

The Apparition tensed, and said, 'Fort Bragg, North Carolina.'

'Back East?'

'Yeah.'

'Were you needed there long?'

'Oh, a few years. I really liked it there.'

'Look, let's stop kidding each other,' I said. 'We both know that's where military intelligence operations are headquartered.'

'Gee, I wouldn't know anything about that. Say, you guys looking for something special here?'

'Vultures,' I told him.

'What?'

'We came to see the vultures.' I don't know why I said it, exactly. Something told me, if you invoke the bird of death, the Apparition will disappear.

'That's really gross and disgusting, you know,' he said.

'Is it? By the way, could you direct us to the slaughterhouse? That's where the vultures usually congregate. Or a place where there's rotting flesh. Either one will do.'

With that, the Apparition vanished.

The fine old adobe church was shut tight and guarded by soldiers, but behind the town rose a long, steep hill, and at its top was a neat, white sanctuary, inaccessible by car. I left Mickey, who wanted to find the market-place, and huffed and puffed the mile uphill. The climb had evidently discouraged the soldiers, because the sanctuary remained nominally open, though effectively deserted. The stall outside the courtyard, which sold devotional candles and sodas to pilgrims, was closed. A peasant boy was grazing his *burro* on the fresh green grass growing up around the building. There wasn't a priest in sight – only the old Indian janitor in his stiff white shirt buttoned to the neck, pushing his dry mop up and back as though absorbed in a crucial spiritual undertaking. Maybe he was, at that.

'Come in, come in,' he said. I had hesitated at the entrance, not wanting to dirty his clean floor with my mud-caked boots. The black-and-white floor tiles gleamed spick-and-span under the mild afternoon sun beaming through the stained-glass windows. But the man sweeping the floor said, 'Come in. Don't worry about it. You're welcome here.'

The simple wooden pews had been pushed against the walls, leaving a wide, polished communion lane, the apple of the floor-sweeper's eyes. It led to five tables in front, each covered with white lace embroidered cloths, alight with burning devotional candles. Behind them, the altar had been freshly decorated with long-stemmed lilies, gladioluses, and baby's breath. The polychromed crucifix peeked from behind the bouquets: Jesus' body fully wrapped in a yellow woven shroud. I suppose the Indians were too modest for a naked Jesus, even on the cross. The walls of the sanctuary displayed the bright circus colours the Mayans have always favoured – pink, green, and brown stripes circling three sides, decorated with hand-painted medallions. Wooden arches supported the ceiling, and the boards between the struts were painted in the same black-and-white squares as the floor. Between this chequerboard ceiling and the striped walls were five murals in that same lively Mayan style. One showed the baptism of Jesus, another the first baptism of the Indians in this parish on 2 July 1545, at the hands of one of Las Casas's black-frocked monks. He was portrayed as a roly-poly man with a big black beard. There was a Quetzal in attendance to witness the conversions, winging high near the sun. The rear wall near the entrance, however, was more representative of the contemporary religious scene in the highlands: it held hundreds of snapshots of disappeared persons, mostly internal passport identification photos. They were pinned or taped to the wall, and relatives had scrawled messages on scraps of paper – a name, the date and place of their relative's disappearance, or a prayer to the Virgin for their safe return. They were mainly photos of teenagers, Indian boys and girls, poor kids with rotten teeth, scared looks, flash-bulb lights in their eyes.

I came out into the fresh sunlight to admire the sanctuary's façade, red, green, and white, with false columns and false scrolls, and double bell towers with a cross perched above a centre-arch window. A teenage couple cuddled on a small bench at the edge of the courtyard – something rarely seen in Guatemala, where even youthful romance seems to have suffered repression. The pair fled before me like flushed quail. From the lookout they had occupied opened a wide alpine view, green furrowed mountains, ochre valley. An immense spiral of Zopilotes circled over the town. There was indeed a slaughterhouse down there. It was drawing vulture activity like crazy.

Walking back downhill, I found Mickey madly bargaining in the market-place. He was following the precepts of Tomás of Chichicastenango, heaping insults on a *huipil* with orange, lavender, red, and green embroidered flowers. The lovely young Mayan girl who'd taken up his challenge laughed wildly at his antics. Her more pragmatic mother made ready to close the shop as soon as her daughter closed the deal. When I finally dragged him away, we drove out of town, stopping at the slaughterhouse to check out the Z-birds. There were a number of concrete holding pens and a concrete killing floor, smelling of sour pigs' blood and disinfectant, all under one long metal corrugated roof. The crest of the roof was strung with stupefied Black Vultures like gargoyles. The offal, mostly manure with a few entrails, was dumped in mounds behind the building, where the chickens and a few vultures still feeding peaceably shared the turf.

The butcher, García, stood in a smoky little cookhouse in his bloody hip boots and sweaty undershirt, making himself something to eat on a small charcoal brazier. 'Sure, go ahead, take as many pictures as you like,' he said with refreshing openness. 'They're almost like pets here, the Zopilotes. We start slaughtering early, about 2 a.m. because the meat has to reach market by daybreak or it starts to go bad. The Zopilotes start arriving about dawn. By seven or eight [a.m.] there's usually a hundred or so collected in the yard and waiting in the trees for the boys to bring the shit outside. Isn't that right, boys?'

The boys nodded lazy agreement from the kitchen floor, where they were trying to sleep.

'They eat anything, but especially they like the shit,' García went on. 'And they're smart, too.'

'How can you tell?'

'I've been at this job for thirteen years now, and you know what? Look, we slaughter only on Thursdays and Saturdays, because those are the market days, right? Those Zopilotes, they don't show up on any other days but Thursdays and Saturdays. I believe they know what day it is.'

'They clean up the yard for you, eh?'

'*Ah, buenísimo!*' he cackled. 'Look, in all the years I've been here, we've always had just about the same number of Zopilotes. They haven't really increased in numbers, and they haven't diminished. So, they never go hungry, because there's never too many of them, see? That's smart, right? They come and eat in the morning, they take a little sleep, then they go up into the mountains for the night, where they have their nests in the caves. But no man ever sees the Zopilotes' nests, because they won't let men see them. Only the children can find them. Isn't that right, boys?'

The boys were sleeping and didn't bother to nod.

'Yeah,' said García. 'Life is good for the Zopilotes. I admire them a lot. They live a long time, I think. About seven hundred years. You can really grow fat eating all the shit around here.'

18

Finding the Dary residence in Guatemala City did not prove easy. There was a phone number in the directory, but I couldn't get the pay phones at the airport to work. When we found a functioning phone in town, the woman who answered was not Dary's widow, but some relative of the immediate family. After a lot of suspicious questions, she gave me a telephone number and street address. But no one answered the phone there, so we just drove out to the house. It was in a posh neighbourhood, where armed guards were posted in front of the residences, toting Galils and Uzis. The bougainvillea bloomed thick, the walls were high, and the Chevy Blazers wore shades. The grass beside the sidewalks looked as perfectly manicured as a golf green. The only people on the street were uniformed maids, gardeners, and chauffeurs. Strands of barbed wire had been strung along the top of the adobe walls, and jagged pieces of broken glass bottles were set into the top of the perimeter walls, so that no one could climb over without slashing himself to ribbons. It looked like an attack was expected any day now.

At the very end of the street, the Dary house stood unprotected. With its wrought-iron gate, flower beds, driveway, and carport, it looked like any home in an affluent suburb of Miami or Los Angeles. The only thing standing guard there was a hummingbird, sipping nectar from the showy red hibiscus blooms. It was a Little Hermit Hummingbird, all bronze and tan, with a long, curving sword bill. It perched for a minute or so, shamelessly staring at us, as if we were the first humans it had ever encountered. A maid came down to the gate at the ringing of the bell, trailed by a handsome little brown-haired boy in short blue pants and a clean white sports shirt. His face was full of anxiety, and after the maid went back inside the house to announce us, he aimed a long wooden pole at us like a spear and heaved it as hard as he could. But it was twice his own size. The spear rattled against the iron gate and fell harmlessly to the drive. The boy picked it up, retreated a few steps, and launched it again. I reckoned he was the son of the murdered Mario Dary.

Eventually, a young woman of perhaps twenty came out and told us to pass through the gate. She showed us into the living room and brought glasses of cold lemonade. 'How can I help you?' she asked.

'You are the family of Licenciado Mario Dary?'

'Yes. I am his daughter, Claudia.'

'I telephoned another Dary, the only one in the book, and the woman gave us directions here. She said she was family.'

'My aunt,' said Claudia. 'Ours is a very small family. Everyone knows where we are. Unfortunately, my mother isn't home . . .'

I explained our visit to her – simply to know something more of her father, his work for the conservation of the Quetzal, and the circumstances of his death. She listened thoughtfully and without emotion, and I realized she was the first woman I'd been able to speak to in Guatemala on any subject more serious than a restaurant bill. She was not one of those haughty, spoiled children of the oligarchy, flashy, dolled up, and inevitably superficial. Claudia Dary impressed me more as a grown-up bookworm: I could see her at age seven or eight, crawling into her daddy's lap with her big, thick eyeglasses and a book of animals in her hand. She still wore those big, thick glasses, and her long dark hair was tied in girlish braids. She seemed very vulnerable and shy as she began to talk of her father in a musical voice, much like the Mayan voices we'd heard in the highlands – I wasn't surprised to find out later in the conversation that Claudia was a graduate student in anthropology, specializing in the oral folklore of the Guatemalan Indians.

'My father became rector of the University of San Carlos in June 1981,' she began. 'He wrote and lectured a great deal, not only on the Quetzal, but on other species and conservation topics as well. We're just now trying to gather his papers for publication, but it's difficult. There are so few interested here in conservation and the environment. Since my father's death, the Biotopo system, too, where you have been, has suffered. The land itself there belongs to the university, but INGUAT, the state tourism organization, has a great deal of input, as they administer the Biotopos. It's not good, because conservation and tourism can work at cross-purposes. INGUAT wants to bring more and more people to the Biotopo, but what the Quetzals need is a reserve where there aren't any people. For example, my father proposed that the Biotopo be closed to visitors in the breeding months of April, May, and June, so that nidification could take place without disturbance. But so few understand that if our national bird is to survive, we must take account of the bird's needs, too, not only our human needs . . .'

Her voice trailed off: sad truth is the stock in trade of the Central American intelligentsia. We finished our lemonade in awkward silence. Then I said, 'We heard in the highlands that your father was murdered – that is, assassinated. But no one there knew anything about it. For example, who killed him, and with what motive? Were his murderers ever arrested and charged?'

'No,' she said. 'There are many criminal gangs here in Guatemala – "máfias", they are called. We believe one of these gangs assassinated my father. You see, the University of San Carlos is an autonomous institution, according to its original charter, given by the Spanish Crown to the Church in colonial days. This means that the authorities cannot intervene in affairs on campus. The police cannot invade the university. During the 1970s, drug dealers located themselves and their rings at the university, to take advantage of this protection. Since the campus was off-limits to the authorities, they could operate with impunity there. The violence had become rampant on campus when my father became rector. He was very much against this. He said in public that these "*máfias*" were destroying the academic atmosphere. The students could not concentrate on their studies, there was so much violence, so many guns and criminal acts. He announced that he was going to get rid of these "*máfias*", even if it meant breaking with the tradition of autonomy by allowing the authorities on to the campus, because the university could not tolerate the murders and the drugs.

'We believe that these gangs hired an assassin to kill my father. It's not hard to do in Guatemala, with a little money. You can hire a murderer for forty quetzals. My father was in great danger from the "*máfias*" because of his strong decision to eject them from the university, but he insisted that he needed no protection. He never carried a weapon, he wouldn't have any guards around him. This is just the way he was. On 16 December 1981, exactly six months after he became rector, he left his office as he did every afternoon. He was walking to his car when they shot him with a revolver; I don't know the calibre or the make. He died before reaching the hospital. An ultra-radical political group that no one had ever heard of claimed credit. But whether they were actually responsible, or were the hired ones, or were just trying to gain publicity, or were a rightist group trying to put the blame on the left, we never found out.'

'There was no investigation?'

'Yes, not just then, because the former government didn't do anything. But immediately after the coup that put Ríos Montt in, a police investigator called to say they were making an investigation and had some suspects. This was when the regime was trying to give the appearance that things had changed, that the government was going to get the "*máfias*" and the death squads under control. Because at that time there were bodies in the streets every day. People were afraid to come out of their houses.'

'Did anything come of the investigation?'

'No, nothing at all,' said Claudia Dary. 'That's not so unusual here. We have many, many unsolved murders. We have many unpunished crimes. That is the history of our country.'

Before we left, Claudia went upstairs to her late father's study and returned with a paperback published by the university containing

Dary's discourses during his short tenure as rector. I had the chance to study them during our return drive into the highlands. They were the speeches of a humane and liberal man, but more so of a scholar and educator, intent on warning his countrymen of its dangerous situation. In a lecture entitled 'The Quetzal: The Guatemalan National Bird as a Symbol of Conservation', delivered two months before his death, Dary had denounced the attitude and policies of the Guatemalan government in the field of conservation. 'National parks' hardly worthy of the name had been created willy-nilly by decree in 1955, he said, then all but abandoned to squatters, poachers, timber cutters, scavengers, cattle barons, and Mayan temple robbers. Administration of natural resources had been handed over to institutions like the Forestry Institute, and bureaucrats like the tourism officials at INGUAT, whose principal interest was to exploit, not to conserve, to profit from nature, instead of protecting it. In a region noted for its poverty, Guatemala was still the only country in Central America that had not committed itself to the rescue of rapidly depleting cloud forest, the Quetzal's only possible habitat. Even poor Honduras, war-torn Nicaragua, and El Salvador had made more effort. It was not until the University of San Carlos formally approved the Biotopo Universitario project in 1976 that the state even recognized the problem.

Dary called on his countrymen to pay serious attention to the conservation of Guatemala's natural birthright before it was too late. A national effort was necessary because Central American conservation organizations aren't strong enough, well financed enough, to create sufficient biological reserves through private initiative. Out of Guatemala's total area of 108,899 square kilometres, Dary said, the Quetzal had originally been distributed over 25,000–30,000 square kilometres. By 1974, that range had been reduced to 3,500 square kilometres, a tenth part. By 1981, the Quetzal's zone had been further reduced to 2,500 square kilometres. Yet national parks, he declared, have come to constitute 'a vital necessity for the survival of the human race'. Without them, Guatemala would continue to lose not only its trees and birds, but its soil, water, air quality, and the remaining fish, reptiles, and insects. An ecological crisis of such proportions would wipe out any chance of social peace and balanced economic development.

Dary was blunt: without rational environmental planning, informed by ecological science, there was no hope for the future. As the Quetzal goes, so goes the Guatemalan nation. 'Unfortunately,' he concluded, 'man cannot create species, he can't create nature, he can only destroy it all . . . The Quetzal cannot continue as a theoretical symbol, because it is a concrete reality. It lives, and continues to exist, for the pride of our country.'

IV

The Last Boss of Chelem-ha

I

    Driving towards Chelem-ha in the coffee country
south of Cobán we could see the alterations the rainy season was
making in the Guatemalan highlands. On roadside *milpas* the maize
popped up in a robust green hue. And scrubbed by the downpours,
the mountain ridges turned a velvety golden green under a drying sun,
which peeked out now and then between racing clouds. The air was
so clear at 1,500 metres that we could watch the Río Polochic wind
like a great green serpent through the valleys, down to its emptying
place in Lake Isabal. The streams and tributaries gushed prodigiously,
carrying tons of khaki soil, romping across roads and pastures in their
zeal to join the main branch of the river. Our progress was slow. We
had told Alfredo Schlehauf that we would arrive at Finca Remedios,
his coffee plantation, that day, leaving the Toyota four-wheel in his
keeping before trekking on foot to Chelem-ha, the mountain he had
called the last wild refuge of the Resplendent Quetzal.

As we climbed from the basin of the Río Polochic, the road grew
worse, but the weather got bright and intensely humid. Butterflies
rose up in waves, in onslaughts – so many that if they'd been money,
the poor people of Central America could have paid their debts in a
single afternoon, with enough left over for a big drunken fiesta. The
giant chameleons came out on the road to warm their innards, marching along with such a brisk step-and-slither that their skins blushed
lime green, and the dewlaps under their chins pulsated salaciously like
neon signs. We roared through several small towns, stopping only to
say hello to the Guardias, mostly young boys on this Sunday, passed
out in sleep on the sandbag barricades, embracing their fathers' rifles.
In Tamahú the people had come out to the centre of town, and in lieu
of having a real plaza of their own, they danced to guitars in the
imaginary plaza of the crossroads. On a grassy field outside Tucurú,
the young Indian men were playing soccer: one was surprised to see

their team uniforms *not* woven of fabulously coloured threads, for the intensity of the game was reminiscent of a Mayan ceremonial ball game. And from all the many roadside churches, wooden sheds erected in thickets of vines, wild pansies, and yellow trumpeter flowers, came the waning, wandering voices of the evangelical congregations, singing their Sunday hymns. Our clothes stuck to our skins like paste from the cumulative damp. Past the towns and into the foothills we stopped to air our feathers, admiring the immense Lobster Claw and Bird of Paradise flowers, which grew in such profusion along the way. The mud changed to dust, the dust back to mud, and the river road, washed out over numerous stretches, had to be forded slowly in four-wheel drive, water lapping against the bumpers.

The pace slowed again after Papaljá, a banana depot burned to the ground lately in a guerrilla raid. The broken pavement ended, and a one-lane dirt track coiled back upon itself precipitously through polished green banana groves. With all four wheels engaged, we bumped and weaved up the 3,000-metre bowl. The earth below was revealed in such dizzying visions that even we, jaded by the unearthly natural beauty of Guatemala in previous weeks, went against better judgement, and stopped to join a lone Indian, gazing in a silent trance at the mountain tiers as they vibrated under the conical hats of the clouds.

But stopping proved a reckless idea; a few kilometres above the lookout, the road petered out in a thicket of high yellow bamboo. Under the spell of the highlands, we had taken a wrong turn, and would have to backtrack. The road down didn't turn out the same as the road up. And asking directions was, of course, no use. To our increasingly desperate 'Finca Remedios?' the helpful Indians who came out from their roadside huts keened, '*Ahhhyyybuen,*' and went on to explain our position at great length, in Kekchí. Their voices floated like wind chimes on the rarefied mountain air, but they may as well have been discussing the meaning of the morning star to their Mayan forefathers. We were royally lost. Then the sun surrendered, and everything round suddenly turned grey; mountains, trees, and sky. A fine mist carried the Toyota up into a sketchy realm of coffee plantations fretted with barbed-wire barriers, where armed guards came out from behind the trees and listened with drunken, basilisk eyes to our claim of getting lost. Only hours later did we find our saviour in a good-natured Ladino, welding a truck in the shed of a *finca*. He strode through the fog towards the car in his welding goggles, the brazing torch still in hand – a mythical warrior from an age or planet you couldn't quite determine.

'Excuse me, *señor*, but we're a little lost.'

'Where are you trying to go?'

'Finca Remedios.'

'In that case you're more than a little lost, you're a lot lost,' he laughed. 'Better let me draw you a map.'

The last dregs of daylight were sliding into the nooks of the trees. We simply didn't know any more if we were advancing on a road, through a *milpa*, or along the path of a rivulet, when a clearing opened ahead, and we finally pulled up beside the Finca Remedios farmhouse.

It had taken six hours to go the last thirty kilometres.

The burly figure of Don Alfredo came round the side of the house and greeted us with a wave. He was clad in a white undershirt, old green pants, and muddy work boots. 'I was beginning to think you decided not to come,' he said.

'We were beginning to think we'd never make it.'

'My children left in the truck hours ago to make the Cobán Road before dark. You should have passed them on the way up – but no, you must have had some problems with the directions.'

'Some.'

'Come on.'

He led us around to the unscreened porch behind the house and offered us – what else? – a cup of coffee. 'I apologize in advance: it's not our own coffee.' He pushed his thick eyeglasses thoughtfully back up his nose and spoke in his friendly Billygoat Gruff voice. 'I can't afford to drink my own coffee here any more. It's all for export. But this is not bad coffee. The water here is pure, good for making coffee.'

'You drink a lot of coffee, I take it,' I said.

'Well, you'll see. If only we had an electric coffeepot.'

'We would have brought you one if we'd known.'

'It wouldn't have been any use. The power lines were cut last year. Sabotage. Anyway, you'll stay the night and we'll see what the weather looks like tomorrow for going up to Chelem-ha. The accommodations here are very simple, rustic – but you must be used to camping, if you've been going all over the countryside looking for Quetzals.'

With that he went into the kitchen and conferred with his Indian housekeeper in Kekchí, which Don Alfredo speaks fluently. Back and forth, they sounded like a soothing oboe duet played on a soft summer night. You could have sworn only lovers could talk to each other in such gentle tones. When they came back out to the porch together, our host was carrying a beat-up, two-gallon tin coffeepot. The dark gnarled woman, whom I could hardly make out on the unlit porch, set bowls of beans, scrambled eggs, and green chillies before us. She was holding a stack of small *tortillas* in the crook of her elbow. 'You must be hungry after your long trip,' Don Alfredo explained. 'But I warn you, be careful of these chillies. *Pican mucho* – they have a big bite!'

He lit a candle and joined us. In the frail yellow glow, we sat on the hard benches, safe from the rain, which had begun to descend in light but steady sheets. The roof lacked gutters, and the water cascaded freely down the grooves of the corrugated metal, forming a beaded curtain in front of our eyes. I asked to use the bathroom and excused myself. Inside, the house had seen better days. The big mahogany doors were being assaulted by some of Central America's thousands of wood-eating insects. Algae and moss were climbing up the cracked and peeling plaster walls (but no orchids). The unadorned second-floor veranda, jutting into the dark and rainy night, reminded me of a Mayan burial mound standing undiscovered in the jungle. A pair of austere nineteenth-century portraits of the German ancestors hung funereally on the living-room wall. The damp chill of the highland evening wheezed through everywhere; there was no heat. Yet we felt cosy enough around the back porch picnic table, laying on the mortar of black beans to our stomachs, and surrounded by the non-functioning crank-up telephones, the mildewed account books, the high, windowed cabinet which protects the plantation's medicines from the climate. Don Alfredo Schlehauf was obviously a man who had made his peace with rural solitude – his wife and children live in Guatemala City, hundreds of kilometres away, while he runs the *finca* alone, commuting on weekends, when they come for a visit. For the past two years, however, the danger of violence in the countryside has prevented their staying overnight. For nearly a year Don Alfredo himself had been warned to stay off his *finca* – there'd been death threats against the local landowners, still viewed by some as foreigners because of their German or Swiss extraction. As a result, last year's coffee harvest was entirely lost. His several hundred workers lost a year's wages. Then his teenage son, who takes no interest in agriculture, totalled the family car. Schlehauf had given his children the old *finca* truck to drive back to the city. Now that he had returned to work at the *finca*, he had no wheels. He had to spend all day each Friday walking down the route we'd just traversed by car to catch the bus for Cobán that runs once a day, transferring there for the long-distance bus to Guatemala City. On Sundays he reversed the entire journey. Only a hopelessly addicted countryman would do that and not feel embittered. Yet we found him in high spirits and exceptional appetite, blatantly relieved to be back in his tropical manse, in his muddy boots, several days unshaven, singing in Kekchí to his ancient housekeeper. He seemed genuinely pleased to have our company, and listened enthusiastically to our tales of Quetzal-watching in the Biotopo sanctuary of lower Verapaz. He was especially interested to learn about the experiments with artificial nesting holes.

'I've thought of trying something similar at Chelem-ha, and maybe I will, some day,' he said. 'The dead tree trunks where the Quetzals make their nests have become scarce. I made a home movie of the way the birds widen an old woodpecker hole. They enlarge it so the male can go in and out without damaging his tail plumes, but they also deepen the hole and make it recessed, so their eggs actually sit down in a cavity, where they are safer from predators attacking by air. The predator can't reach down far enough to get at the eggs. It's the only real protection the Quetzal's eggs get, because I don't think the parents defend their nests if directly attacked. Snakes can still reach the Quetzal's eggs, of course. But at Chelem-ha, at least, it's too cold and rainy for big snakes to be much of a problem.'

Mickey had been following our conversation in Spanish with drooping eyelids. Now he seized on the words 'snakes', 'cold', and 'rainy'. His head receded with a shiver into the woollen shoulders of his sailor's sweater, and he flashed me a dirty look. Out of the rainstorm beyond the porch appeared one of Don Alfredo's drenched farmhands, barefoot, in a nylon poncho: his wife had taken sick, and the man wondered if *El Patrón*, as Don Alfredo is called, had any medicine. Schlehauf dispensed aspirin and liquid antibiotics from the porch cabinet, told the man what to give her when, and said to come back the following day if there was no improvement.

'Yes, one of the problems in the conservation of the Quetzal is this lack of tree stumps for their holes,' he continued to muse when the Indian had gone. Another round of the strong coffee was poured and sweetened with raw sugar. 'The trunks the Quetzals choose tend to be old and rotten, since their soft bills can't penetrate hard wood. Some trunks collapse, others the people destroy. A few years ago, one of the men who lives up on the mountain came to me one day and said, since he knew that I take an interest in birds, that he'd found a Quetzal nest in the trunk of a dead tree. I questioned him on the size of the nest, the eggs, and so forth. He seemed to know what he was talking about. So he went up to work in his *milpa* and I went to Chelem-ha to take a look. It must have been in, let's see, March or April, because the male Quetzals were making their nuptial flights. This is something to see! The *macho* flies straight down the mountainside with an undulating movement, so that the long tail feathers stream and sway behind it.'

'Like a flying serpent?' I interrupted.

'Yes, that's it. Much as you'd imagine a serpent doing, if snakes could fly. Incredible. And all the time it's performing this flight, the male cries sharply, "tak-teek . . . tak-teek" . . . like that. When I got to the place the Indian had described, I saw that what he hadn't told me was that he'd hacked out steps in the trunk with his machete, and his small son, wanting to see the Quetzal nest, climbed up. The rotten trunk collapsed; of course, the nest was destroyed. It was only fortunate the boy didn't get seriously injured in the fall.' He drank his coffee at a gulp and waved his hand in the air casually. 'Here *La Gente* – the people – destroy everything, natural and man-made. They destroyed the electric wires, the phone wires that used to connect us to the next *finca*. Everything . . . everything. If their *milpa* goes up to a certain tree line at the edge of the forest, and they want to expand their *milpa* beyond that line, they just say, "There, that's my *milpa*," and go ahead and burn the forest. They hunt every animal that moves. When they're hungry, they'll even eat armadillos.'

The rain continued to pummel the roof sullenly. The insects zoomed through the candlelight, fell with singed wings on to the table, and mounded up there like the sands of a broken hourglass: we could barely hold our eyes open. 'Well, enough for tonight, you're exhausted,' said Don Alfredo at last. 'Let's just take a last coffee, and I'll show you upstairs where you'll be sleeping in the children's bunk-room.'

'It doesn't keep you awake?'

'What, the coffee?' he asked, with the baffled look of someone who had never encountered the queer notion that caffeine is a stimulant.

2

In the morning the steady drenching rain continued. Smoky panels of fog crept through the arbours surrounding the *finca*, allowing only occasional, blurred glimpses of the clouded mountain peaks all around: blue ranks of levitating, tufted spikes. Over a pro-longed breakfast of beans, cream, hot chillies, new *tortillas*, and fresh coffee, the Don advised against starting out for Chelem-ha – an eight-kilometre climb up a poor dirt road, impassable by car in the rainy season, then a further hike of twelve klicks on a muddy foot trail to the base of the mountain ridge. 'It's a full day's journey on horseback,' he said, adding seriously, 'that is, in good weather – and if you know the way.'

'Tech meeting!' blurted out Mickey. We conferred. Overnight he had decided against going up to Chelem-ha, and Schlehauf's road report clinched the matter. He envisioned sitting around Finca Re-medios for days in wet clothes, waiting for a break in the weather, only to undergo a long, miserable, and probably fruitless march into the misty mountains. From a photographer's point of view, he was absolutely right. Our visas would be up in a week; in the limited time left, our tasks had become different. It would be more productive for him to try to get pictures of the small flock of Quetzals we'd become familiar with at the Biotopo in Baja Verapaz. I still wanted to check things out here in Alta Verapaz. Besides, I was also growing interested in Don Alfredo, a man who seemed, in his manner and conversation, to be engaged in an unpublicized wrestling match with himself –

perhaps his liberal sympathies and conservative values struggling to gain the upper hand. He seemed a remarkably reasonable man; that alone was sufficient to arouse one's interest in such an unreasonable country. In any case, Mickey and I had come to a professional parting of the ways. No question of desertion.

We agreed to meet on the Cobán Road in five days' time, and after breakfast, Mickey said goodbye to Don Alfredo, who understood the situation, yet accepted it with regrets.

The tail lights of the Toyota four-wheel dissolved into the fog.

A short time later, the field hands showed up at the back porch to find out if *El Patrón* wanted them to work in the inclement weather. Don Alfredo bantered with them as he laced up his work boots, then went off to attend to the cultivation of the coffee bushes. I didn't see him again until mid-afternoon, when he returned for a dinner of leftover beans, *tortillas*, and an even more biting chilli, which made the coffee taste like firewater. Smoking his after-dinner cigarette, Don Alfredo said, 'Well, the weather has broken for a while, but it's too late to start for Chelem-ha today. If you like we can ride up together, if it's good in the morning. I gave my horse to one of the people. I'll send for him, and he can get another horse for you. I haven't been up to Chelem-ha since the rains started, and besides, I have some business to attend to there. Why don't we go out to see the birds on the *finca* this afternoon?' He added with an ironic twinkle in his eyes, 'You can identify them for me. I've always wondered what their proper names are.'

'I'm sure you don't need a gringo to identify the birds on your own *finca*, Don Alfredo.'

'Really, I only know them by their local names, the names the people make up for them,' he protested mildly. 'You know, I've never owned an identification guide. I'm not sure one even exists in Spanish. For instance, I'd like you to see one of our birds here, very colourful and interesting, we call *pájaro reloj*.'

' "Clock bird," ' I translated. 'Some sort of cuckoo perhaps?'

'Koo-koo,' he repeated the syllables, then shook his head. 'I don't think so. The *pájaro reloj* gives a low-pitched call, almost a whistle – "hoot-hoot-hoot" – three times in rapid succession. "Hoot-hoot-hoot". The name comes from the way it moves its tail, like the pendulum of a clock. Anyway, let's go see.'

The day had lightened to the best of its puny ability as we entered the drizzling woods. In the shade of half-wild plantains and thatch palms, Schlehauf had planted an acre or so of cardamom, an Asian ginger growing in tall, leafy shoots, propagated with great success in Guatemala in recent years. The hard, dark, tiny fruits are exported to spice the perfumes of Europe, the coffees of Arabia, the baked goods of Scandinavia. Beyond lay the twin silted ponds of the *finca*'s water system. A swaggering male Ringed Kingfisher chattered and dived for the overgrown carp swimming in thick schools near the surface. '*Martín pescador* – Martin the Fisherman – he's called,' whispered Schlehauf, and trooped off round the pond to drive the bird towards me for a better look. The kingfisher played hide-and-seek in the low branches near the water's edge, always giving its position away by dint of its dirty red breast and wide white throat band. In the process of flushing the kingfisher, Don Alfredo also raised a Band-backed Wren, several Clay-coloured Robins, and then two of the more exotically feathered local residents – a Flame-orange Oriole with its breast spotted black like a pattern of gunshot round the heart, and a Crimson-collared Tanager, all silky crimson with a jet-black face mask.

'Excellent birding here,' I enthused when we joined up again at the far side of the ponds.

'I've tried to leave the old trees standing so the birds will have some cover,' Don Alfredo said. 'A few years ago I stocked the pond, thinking we would eat the fish. I never imagined they would grow so large. They were only as big as my thumb then. Now even *Martín pescador* has trouble carrying them away. It's a shame there's no one here to eat them. But what the hell – the people will steal them soon enough, eh? But I wonder where the *pájaro reloj* is hiding. He's usually around calling in the woods this time of day.' He led up the footpath, half-whistling his 'hoot-hoot-hoot ... hoot-hoot-hoot ...' To which a handsome Mottled Owl, roosting in the canopy, suddenly woke up and started hooting in response.

By half-past four we had reached the rocky scrub summit of the humpbacked mountain rising behind the *finca* farmhouse. White-collared Swifts, twice the size of North American swifts, breezed around the crags in effortless circles, figures-of-eight, loop-the-loops. We sat for a moment to take in the vast alpine extent of Central

America's mountainous backbone. 'Over there,' Don Alfredo pointed out a barren brown patch. 'You can see the spot where the *beneficio* used to be. It was a government-run farm that raised animals and plants for distribution to the people. The rebels burned it in the raid of two years ago. It was a stupid thing to do. I think it turned the people against them. The *beneficio* was the only thing the government had ever done for the people here, the only place the people could obtain turkeys and pigs to raise. And it was a nationalized farm – the rebels ought to be in favour of that. How can I understand an act of destruction like that? I'm not well informed about politics, but I don't think the rebels really know what they are doing. It was after that raid that some of the landowners in the area fled.'

'Maybe that's what the rebels did want – for the landowners to leave.'

'Maybe. Who knows?' He spoke, as usual, with neither anger nor mockery. 'They never made any demands, only threats. Sometimes it seems to me we have a civil war whose only point is for both sides to destroy each other and the whole country, too.'

'But you came back. What happened?'

'The army came and caught the rebel leaders and killed them. That is what happened.'

The Don grew pensive, but not at the thought of the rebellion in the highlands, which has been four hundred years in the making and which he can have little influence over in any case. It was something more personal, he said: his first daughter was buried in a rock tomb close by, up here where the clouds and mountains met. I stammered my condolences. 'It's a good place, magnificent view,' he went on graciously. 'She died at birth. It was many years ago, when we first bought the *finca* in 1958.'

I reacted with surprise, for it had seemed to me, from the shambling old house, the dusty portraits of the forebears, even from the swarthy café colour of Don Alfredo's skin, that Finca Remedios must have been in the Schlehauf family for several generations. 'No,' he said. 'My brother and I bought it together in fifty-eight. Before the Second World War my family had another *finca*, near Cobán. But during the war the North American government got the Guatemalan government to expropriate lands belonging to Germans, and even interned some of the Germans in Texas. They suspected them of Nazi sympathies.

And there were some – the Germans in Guatemala have always been nationalistic towards Germany and resisted assimilation into the population. Like most colonies of Europeans in foreign countries, I guess, only more so. My father was German by birth, my mother Guatemalan of mixed race. I'm the first generation of my father's family born in Guatemala, the first to say, "I'm Guatemalan, not German." So we started again in fifty-eight. But only a few years after we got our coffee trees to the state of production, the conflict and the violence started. Then coffee prices fell. So things haven't been going well. When you're a farmer, you just say to yourself, "Maybe next year will be better" . . .'

'Or you sacrifice to the gods.'

'Maybe if things get any worse, we'll have to try that.'

Presently, when he was satisfied that the grave site was in good order, we threaded our way back downhill to find the *pájaro reloj* before dusk locked in. The clouds were already on a low roll, carrying the coming night's rain. But the Don wouldn't be satisfied till he'd found the *pájaro reloj*, so luckily one finally put in an appearance, signalled by a low-pitched 'hoot-hoot-hoot' in the undergrowth. '*Ayy*, there you are, *pájaro reloj*. Hoot-hoot-hoot,' my host said, more to the bird than to me. He stood motionless, one finger hoisted in the air. The bird responded. Then Don Alfredo got down on his hands and knees, and began to crawl through the barely penetrable thickets. I did not want to damp his enthusiasm when, after a wet chase, his *pájaro reloj* turned out to be none other than the Turquoise-browed Motmot, the phosphorescent bird with the queer racket tail I'd watched in perfect sunshine on the road up to the Verapazes several weeks and several hundred kilometres ago.

That evening passed in a rainy tranquillity so absolute that even renewed talk of Don Alfredo's problems with preserving the Quetzal's habitat on Chelem-ha could not shake it. Problems – that is what he called them, as coffee followed cigarette following coffee. He laid out the problems one by one. It was like a jigsaw puzzle of Central America itself, where all the pieces were the real facets of various crises – social crisis, population crisis, economic and financial crisis, political crisis, agricultural crisis, and environmental crisis. And he just couldn't see how they would ever fit together. He had purchased the Chelem-ha mountain ridge at the same time as the *finca*, not for

174 · BIRD OF LIFE, BIRD OF DEATH

agricultural development, but specifically to preserve the cloud forest for the Quetzal, as well as the other birds and wildlife there – and for the *orquídeas*, the wild orchids, which, like practically every Guatemalan, rich or poor, Don Alfredo collects passionately. At first, he left the mountain completely alone, though there were already ten Indian families living at that time in a small squatters' settlement at the foot of the canyon. As the people multiplied, they initiated slash-and-burn maize cultivation in the foothills, then on the lower slopes. Intrusion on the forest was well under way by the late 1960s. It was Jorge Ibarra of the National History Museum in Guatemala City who persuaded Schlehauf to seek government assistance in turning Chelem-ha into a national forest area. It was already clear to conservationists in Guatemala that government intervention would be necessary if sufficient habitat were to be maintained for the Quetzal's survival. More than ten years ago, Don Alfredo had offered Chelem-ha to the nation, if the government would only agree to its preservation and the stewardship that entailed. However, the gift was neither accepted nor refused. The Guatemalan government simply never responded. Was it the civil conflict? The financial crisis of falling commodity prices and rising debts? The coups sending government personnel through revolving doors? The crushing weight of corruption in the military, or incoherence in the bureaucracy? Or perhaps the simple lack of interest in conserving the environment, viewed as a tourist gimmick in a country madly dashing for the gringo dollar.

There was no way of knowing: a decade later, Ibarra was still waiting to meet with the right government officials. In the meantime, the ten Indian families had become fifty families, and the *milpas* spread farther up Chelem-ha. Don Alfredo next considered raising funds in the private sector to make the mountain into a fully functioning private wildlife refuge. There could be trails for birdwatchers, horse rentals, a campsite at either end of the ridge, and maybe a small inn, serving simple meals for 'scientific tourists' like myself. An ornithologist from Texas had come as a consultant, and reported several hundred species of birds using the habitat on the proposed refuge site – in addition, of course, to the illustrious Quetzal. But Don Alfredo had been unable to get the project off the ground. Lack of investor interest, to put it mildly: 'The banks, people with money to lend, didn't understand why anyone would want to come all the way out here to *el campo, las montañas*, a place so remote. It seemed the idea of a crazy man. Guatemalans don't have that kind of consciousness of their environment. They don't love nature. I couldn't get anyone the slightest bit interested. So, what to do . . . what to do?'

3

The hunting of Quetzals was outlawed in 1895, when a rococo presidential decree was issued from Cobán in the highlands of Verapaz:

Considering that the hunting of the Quetzal in distinct parts of the nation where this beautiful bird reproduces . . . threatens to completely destroy the species, which would be extremely regrettable not only for the peculiar beauty of the aforementioned bird, but also because it symbolizes the liberty of the fatherland, the President of the Republic absolutely prohibits the hunting of the Quetzal, under penalty of six pesos, or six months in prison.

This was the time of General José María Reina Barrios, who was known as 'Little Reina'. He was chosen heir, and then duly elected, to replace his uncle, the *caudillo* General Justo Rufino Barrios. The *caudillo* was that special form of dictatorship that developed in Latin America, where one man held not only all political power, but military power as well – life and death power over the citizens. As a form, the *caudillo* preceded the now-common *junta*. Uncle Justo was also quite the businessman. He bought a mansion on Fifth Avenue in New York, as well as a million dollars' worth of choice Manhattan real estate, and put it all in his wife's name. This was on his presidential visit to the United States in 1889, during which Barrios negotiated a deal to build railroads in Guatemala with a group of American investors, including former president Ulysses S. Grant. The railroad never got built, but Barrios did all right by the deal. He improved his family's standing further through loyalty to American and European interests. At one point Barrios got involved in a scheme to cede to North

Americans a route for a trans-oceanic canal through Nicaragua. Barrios financed an army to overthrow the government of Nicaragua, but like the Guatemalan railroad, the canal never got built either. In his twelve-year reign he also bankrolled revolutions in Honduras and El Salvador, with a vision of uniting all the Central American nations into a single federation under himself. He made more money in real estate by selling the property next to his mansion in Guatemala City to the gringos to build a new American embassy. The embassy still stands on the same spot. The Barrios mansion has been converted to a hotel, where we stayed.

In Guatemala, the elder Barrios's rule was stern and swift. He promulgated a liberal constitution, providing for separation of church and state, election of independent judges, and a bill of individual rights. Then he decreed forced conscription for the Indians, ex-propriated Church lands, and expelled the Jesuits from the country. Within the first year of his taking power, Barrios had sold the Church lands to members of his own Liberal party and to foreigners, and used the proceeds to establish the Banco Nacional de Guatemala, Central America's first central banking institution. In the highlands, the Indians were brutally removed from their *ejidos*, the communal lands recognized by Spain after the indefatigable efforts of Bartolomé de Las Casas and his brethren. These were sold mainly to Germans for coffee cultivation. Uncle Justo financed the new industry with loans from his new bank.

In the highlands of Verapaz coffee was soon king. The coffee plant, originally brought to the Caribbean by the French from its native Ethiopia in 1723, had spread to Central America by the 1860s. It was found to grow best at elevations of one thousand to fifteen hundred feet. The abundance of moisture on the Pacific slopes and in the central highlands of Central America accorded well with the coffee plants' habits. The nature of coffee cultivation, however, did not accord well with Ladino habits, always oriented towards intrigues, gambles, and instant pay-offs, not sober agronomy. The coffee bush (actually a small tree if allowed to grow, but usually kept heavily pruned to increase the yield of beans and simplify cultivation) takes five to seven years to mature. Picking the red berries is extremely labour intensive, while processing the beans involves complicated skimming, pulping, washing, drying, hulling, polishing, and grading.

Few Ladinos could hack it. With the long-term investment and superior organization required, it was German immigrants to Central America who quickly began to dominate the fledgling coffee industry. By 1900, only one in twenty coffee *fincas* was owned by native-born Guatemalans. Cobán, the Christian settlement founded by Las Casas in 1543, became as German as a Freiburg *Bierstube*. The Hapsburg double eagle flew from atop the *iglesia* San Pedro on Cobán's central plaza. German was the lingua franca. But just as the coffee economy required the importation of 'foreign' capitalist concepts like saving, orderliness, patience, and real estate, it required, too, a large, low-paid, seasonal work force – not slaves, who would become dependent on their masters, but an agrarian proletariat, who would be available when work needed to be done and then get out of the way when no longer required.

Much of the Liberal revolution imposed by General Barrios separated the native population from their protectors in the Church, and compelled them to become such a work force. Their *milpas* in the lower montane regions, where they had moved to escape the Spanish conquistadors, were ideally suited for coffee growing. General Barrios issued orders requiring provincial authorities to supply 'the number of hands to the [coffee] planters that they asked for'. That meant, Indian hands. The army was deployed in the countryside to make sure his orders were carried out. The Indians suffered another spasm of repressive violence, from which they have still not recovered. They were dragged off by the hair, tied up in ropes and chains, to serve as *peones*. From before the sun appeared until after dusk, the Indians picked coffee in the fields of their *patrón*, their boss. At night, to instil orderly work habits, they were locked up in jails. A newspaper article of the time protested,

This is the height of injustice, and not even in times of worst oppression were slaves molested in their hours of rest; they were allowed to give themselves up to repose, the only pleasure that the poor may enjoy. In what times are we living that the authorities of Cobán bring dishonour on themselves by violating the most sacred precept of human liberty, which is the privilege of proletarians in every civilized country? At what moment was slavery reintroduced into Guatemala?

The colonial land tenure system, under which the Church owned lands the Spaniards weren't interested in, and the Indians tilled lands the Church had arranged for them, had tended to accommodate the traditional Mayan agricultural system. Like other New World natives, the Mayans had no concept of private land ownership, nor of wage labour. It would be more accurate to speak of the land owning them – or, better, of an eternal marriage between the farmer and the *milpa* he tended. The *milpa* of classical Mayan times had required the commoner to work approximately ninety days per year for his family's subsistence. The rest of his days were in no sense 'free'. The hereditary Mayan lords appropriated the commoner's surplus time to execute works of architecture, sculpture, literature, and pottery; to build the temple cities, and to support the nobility in their astronomical and ceremonial pursuits. The native peasant was certainly 'exploited' by Mayan civilization, in the Western sense of that term. But that exploitation functioned through entirely different means, and for entirely different ends. The Mayan hierarchy appropriated the commoners' time to carry on the religious rites which propitiated the *Chacs* – gods of heaven and earth – on whose goodwill the commoner's *milpa* was believed to be dependent. The Indian had a sacred duty to raise maize on the *milpa*. There was no money involved. Labour itself was regarded as a form of worship. Working the *milpa*, the Mayans believed, was what kept the sun up in the sky. All social classes strove for that end together. Even post-Conquest prayers show this intense religious concentration on the *milpa*:

O god, my grandfather, my grandmother, god of the hills, god of the valleys, holy god. I make you my offering with all my soul. Be patient with me in what I am doing, my true God and (blessed) virgin. It is needful that you [make] fine, beautiful, all I am going to sow here where I have my work, my cornfield. Watch it for me, guard it for me, let nothing happen to it from the time I sow until I harvest it.

For this reason, the kind of capitalism that swept Guatemala in the Barrios years caused a far more profound rupture than a mere forced modernization of a traditional agricultural system: it was an attack on the Indian's religion, way of life, and conception of himself as someone who grew maize on a certain *milpa*. To be, in this sense, exiled from the *milpa*, was to be separated from the self, to become a shiftless ghost, no longer part of the Mayan weave, no longer quite human. As

Miguel Ángel Asturias put it, 'Crops to eat are the sacred food of man, who was made from maize. Crops for profit are the hunger of man, who was made from maize.'

Thus did General Justo Rufino Barrios turn the churches into banks, the sacred bird into money, spiritual values into profane economic values. In the end he legalized debt peonage, a system under which cash advanced to the Indians accumulated into debts inherited by their children. Where the Mayans had been bound to ancestral lands by the whole spiritual fabric of maize cultivation, a money debt now bound the Indian generations to their *patrón* and his *finca*. The Indians had to accept it. What choice did they have? Those who rebelled were killed as common criminals. Most submitted, and kept smaller *milpas* on inferior lands when they were seasonally unemployed on the coffee *fincas*. Some, however, fled like their ancestors before them into the forests. A new generation lived 'under the trees, under the vines'. Undoubtedly, the number of Indians fleeing into the mountains had a negative effect on the Quetzal population. The refugees began burning the cloud forest. Thousands of birds were slaughtered and sold to skin traders, who smuggled the specimens out of Guatemala to stock the European ornithological cabinets in vogue at the time. It got so bad that Little Reina had to issue a second decree, two years after the first one, beefing up the penalties for destroying the national symbol. It began:

> Inasmuch as the decree of 13 December 1895 absolutely prohibiting the hunting of Quetzals has not produced the effects which had been envisioned in issuing it . . .

4

At Finca Remedios that night I awoke with a start: the rain had stopped. I had grown unused to such silence, and now it was enervating. It occurred to me that I'd never watched the sky so much as I had since coming to Guatemala, yet I hadn't seen a single true sunset the entire time. I'd been pitched headfirst into drowsy fogs, black noons, ghost-filled mists, lightning bolts spraying through electrified grey. I'd grown accustomed to the incessant chant of downpours, to the numbing fear that at last the mountains would finally just collapse, this watercolour would wash away, the stars sputter down into the sea. A clean dawn would be a welcome change. Out of the bunkroom window, a hard white sun perched like a pillow on the silver-blue quilt of the sea. Only months later, recalling that peculiarly empty silence, which seemed to indicate that the earth was sound asleep and didn't want to be disturbed, would it occur to me that it wasn't the sun over the sea at all, but the moon hanging over a 'sea' of stationary clouds. The rainy season had penetrated my consciousness, it was taking place inside my head. All the common and normal groundings in the natural cycle of day and night had inverted. Only then did I finally perceive what power the ancient Mayan priests must have held by gaining precise knowledge of the change of seasons, the movement of the stars, by making an accurate calendar, bringing predictability and order to this waterlogged chaos.

Convinced it was already morning, I jerked my clammy jeans and boots on, and crept downstairs to answer the call of nature – and perhaps take an early walk while the weather held. The bunkroom opened on to a flight of outdoor steps, meeting the front door of the house at the bottom, but the front door was locked. I could hear the snoring slumber of Don Alfredo knocking around inside. Absent-mindedly, I wandered off the concrete slab to take care of business under the trees, under the vines; what difference would my piddling stream of water make in the millions of gallons that had been falling for weeks unending?

To be discreet, however, I trudged a few yards into the forest, noticing how eerily dark it was. A stillborn dawn? It would not be out of keeping. I went down farther, beyond the ponds, and cut in under the canopy, where yesterday I'd bent down to inspect Don Alfredo's cardamom plants. Wind shook the water down from the upper storey. The hulking black trees edged a little closer . . . closer . . .

It was not until I'd finished what I'd gone out to do that I noticed how the trees at the perimeter of vision had encircled me and were pointing their rifles.

'Hey, you're not trees, you're men.'

They were all in colourless ponchos, and they didn't have much to say. A weak flashlight dripped margarine light over my body. I hastily zipped up my jeans, possibly for the last time. It was too comic to be truly frightening. What added to the laugh was the passing thought that if I were shot dead, reports back in the States would be taken far more seriously than the facts of this mishap while out urinating in the jungle warranted. After all, I could just as easily have been bitten by a fer-de-lance, or mauled by a jaguar.

A faint gurgle of whispers was the only indication they were discussing my disposition, but it wasn't hard to guess the alternatives. It started raining again: this is apparently mandatory procedure during a Guatemalan stand-off. They were still discussing the issue, though not in Spanish. I guessed by this, although I could not see their faces, they must be a civil-defence squad on night patrol. It would be best to take a chance on revealing myself as Don Alfredo's guest: one is never adverse to pulling rank in a life-threatening situation. Stiffening up to full gringo stature, I pronounced the words 'Don Alfredo Schlehauf,' 'El Patrón,' and 'Finca Remedios' as loud and clear as I could – half-way between a command and the words of a magic incantation. And then, with a hand gesture towards the farmhouse, I invited the trees that were men to follow me to the front door, where El Patrón would clear the matter up . . .

Don Alfredo was amused when I told him the story over breakfast. 'The next time you come you'll have to learn Kekchí,' he chided. 'It can be very useful here, especially if you insist on wandering around the forest in the middle of the night. But tell me, why didn't you cry out for help?'

It had never occurred to me.

The sky drew as clear as it was likely to get in the rainy season, so we rode out together at mid-morning. We were accompanied by Mateo X, an Indian friend of El Patrón's, who has lived as a squatter on Chelem-ha for many years, and Emilio, Don Alfredo's favourite among the local campesinos, a handsome, strapping young Indian with coal-black eyes and ears of cauliflower sticking out from a shock of straight, black, gourd-cut hair, who speaks excellent Spanish. Don Alfredo was glad to be reunited with his old piebald horse. He rode adequately, but with none of the pretension of the Spanish caballero. I just tried to stay in the saddle and keep the thing moving more or less forwards. I had no whip and no spurs, so, as Mark Twain said, I was reduced to arguing the case. It helped to have Emilio on my side. From the beginning he carried my little rucksack, concocting a charming lie to set me at ease over the issue of servility – that he required some weight on his back for balance walking long distances, while I required *no* weight on my back for handling the horse. When the going got tough, or the horse got ornery, he'd take the reins and pull in front. Or I'd dismount and we'd walk together, discussing the fine points of bachelorhood in the highlands, one of agreeable Emilio's fields of expertise. The animal followed right behind us of its own volition, sticking its hairy muzzle into our armpits, suddenly very interested in everything we said, as if not wanting to miss a word. '*Ayy*, damn horse.' Emilio laughed. 'He thinks we're talking about him.'

For several hours we climbed steadily upward on a dirt road red as modelling clay. We met Don Alfredo's *jornaleros*, or day workers, walking single file down to the coffee groves with their shoulder bags of wool or plastic, their machetes in their hands. They work together in the *cafetal*, the coffee plantation, from dawn to midday, then separate to go up to their *milpas* in the afternoon, to tend their maize, beans, chillies, and squashes. The clouds kept their distance overhead, forming a thick roof but leaving a wide bar of diaphanous sky, grey tinged with blue, green, yellow. The booming roar of a distant waterfall sounded like a far-off battle, but there was no smoke rising anywhere over the dulcet green woods. Past midday, we reached the top of a high pass, and stopped to rest. Through a rocky chute the canyon spread north to Chelem-ha, its east and west flanks as steep, and nearly as high, as the towering central peak itself. Here the muddy road ended, or tried to end: an Indian work gang was out taking advantage of a recent mud slide to extend the road a hundred feet farther towards the canyon. They had no tools for clearing the fallen rocks but their heavy-headed hoes, their machetes, and their own arm muscles. Don Alfredo knew them all by name, and delighted in asking after their families and sharing local gossip. They addressed him as '*Patrón*'. As if in his honour, though it could not have been possible, a boy suddenly appeared from one of the side paths bringing two bottles of warm, fruity soda, which we handed round. I was sorry my partner wasn't there to witness how far soda civilization had spread.

In due time, we mounted up for the next leg of the journey. It took us over the pass on a well-trodden, mud-packed Indian footpath, downhill into the canyon, over a low rise covered with pine, and then across the broad *milpas*. The maize stood as high as the horses' eyes. It had not yet tassled, but this didn't deter the animals, which stopped constantly to uproot stalks and crunch them in their teeth. It was beyond even Emilio's powers to prevent the horses from eating a swath right through the *milpas*, but the peasants at work in the hills didn't take it amiss. They laughed and waved at us with their machetes. Somehow, it neglected to rain, though the sun remained in some meditative seclusion, and Chelem-ha, rising northward, became mottled and forbidding under the running shadows of the clouds. Then the wind and fog began to blow. The mountain was almost completely socked in when we reached Mateo's wattled hut late in the afternoon.

The lodge breathed activity and squalor as we sat outside under the eaves on low stools Mateo's children brought us. There were two strictly separated earthen-floored living areas, one for the women and children, which Indian modesty required we should not enter, and from where you could see the smoke of the cooking hearth rising through the thatched roof. The other room was the men's club, where Mateo and his two grown sons in their twenties slept. Here, in a black darkness there was no getting used to, we would put up for the night before hiking the mountain in the morning. In the bustling dooryard, Mateo's countless younger children joined his turkeys, chickens, piglets, dogs, and cats, in an undifferentiated flock. Their game, their job, the only thing they were doing, was packing down each other's droppings into a hard, slippery, stinking marl. His two maiden daughters, however, were strikingly plump, dark-haired, slightly cross-eyed beauties – as Emilio apparently noticed, judging by the long, sheepish glances he sent whenever one crossed the yard on some domestic chore. Behind the hut, Mateo showed off his rabbit pens, his chicken coop, and the small corral of a single barbed-wire strand where the enterprising man kept two cattle – sturdy mountain hybrids with the shaggy breast locks of buffaloes and the broad, square faces of Spanish breeds.

Don Alfredo and Mateo chattered away together in Kekchí, old friends and former neighbours with many farming matters to catch up on. They were about the same age, the same height, nearly the same build (that of a mountain goat), and shared many of the same rural interests. At one time, Schlehauf had kept a little cabin near by on Chelem-ha, where he would go to relax, watch birds, and collect orchids, in more stable times. But it had been impossible to prevent break-ins, so he and Mateo had dismantled it. The wood was still stored in the rafters of Mateo's place, and Mateo still kept *El Patrón*'s bed there, too, ready for him whenever *El Patrón* wanted to visit. After a trip to Germany, Don Alfredo had brought back apple and cherry seeds he thought might flourish in Chelem-ha's moist, cool climate, and given them to Mateo to plant. Now he was thrilled to see how well the trees were doing, on a ledge behind the homestead where Mateo had sowed them. Mateo reported the trees had produced a hundred kilos of apples, though all had been stolen. The cherry trees were two metres tall. 'For years they haven't prospered, and now they've grown so much!' Don Alfredo exclaimed, stepping up to measure himself against the willowy saplings. 'Two metres. Well, that really does make me happy!'

And what made *El Patrón* happy seemed to please Mateo, too. In a way, the dour, hardworking Indian with the scruffy chin hairs was a lot like the middle-aged landowner: their operations were just on a different scale, and Mateo ran a tighter ship. You couldn't imagine one of his sons totalling the family car, if there were one. His poverty required a far stricter patriarchy, which Don Alfredo respected. It was striking how little the Indians' unadorned life in the highlands had changed from what I'd read of Mayan life five hundred, a thousand, even two thousand years ago. Mateo and his clan were still entirely dependent on their maize crop. His religion was still practised right on his *milpa*, and required neither church nor priest. He only left Chelem-ha once a year to go to town, to trade his surplus harvest for sugar, tallow, matches, and rope.

In the smoky lodge, Mateo's sons served their father and his guests the meal the women had prepared. It consisted of *tortillas* and turkey soup, spiked with a chilli that brought tears even to Don Alfredo's eyes. As for me, one sip was enough. My sinuses burst into flames, and my head swam. I quickly gulped down several cups of water they gave me, heedless of the health risks. Emilio dunked his *tortillas* gingerly into the broth till his bowl was drained, and then, without word or yawn, stretched out on the bench by the table and fell into a deep sleep. I didn't have the vaguest idea what time it was when we went to bed – or, in my case, to hammock. I was beat and sore from the trip on horseback. I was famished from not having eaten. My mouth burned from the chillies. I was shivering from the cold, and couldn't sleep. So I drew myself up into a full fetal position, and lay there hoping there was some old Mayan custom I was unaware of by which the host sent his daughters to keep his guests company. Of course, there wasn't.

5

We spent the entire next day on the mountain called Chelem-ha without observing any Quetzals, and returned to the Indian lodge soaked and chilled to the bone. In fact, we saw few birds of any kind. A yellow-foreheaded warbler or two, and a medium-sized hawk too distant to identify, with bright café-coloured feathers under the wings, drifting south, then disappearing down the canyon. There were only orioles in any significant numbers, apparently a species that moves into a new crop cover after the forest has been burned off. At least half the mountain itself was already gone to fire and axe. The lately felled trees lay sadly in among the maize. Some were slashed down with machetes, others burned first. Charred trunks stretched on the ground beside their charred stumps like defeated totems, surrounded by piles of fresh, wet chips, and the strong mixed scent of pine and charcoal. We passed through a small *aldea* (village), of newly built pine huts. Farther up, a ledge cave Don Alfredo said the guerrillas had used last year. It was so tiny and dingy it could not possibly have sheltered more than three or four truly desperate men, a hideout far more than a base of operations. The military would have a tough time getting in here without helicopters. Above stood the last five hundred metres of virgin cloud forest. Water rushing, dripping, gurgling, bubbling everywhere, and ariot with angel trumpets, blue hydrangeas, lavish tree ferns, mosses, and the orchids Don Alfredo and Mateo stopped to inspect. But no Quetzals. The closest we came was an empty nest hole among some heavy-bearing miniature apple trees, standing out behind an Indian hut at the forest's edge. Don Alfredo rapped on the hut wall and spoke to a woman on the other side, who didn't come out. She said the Quetzals came to feed in the fruit trees at five in the morning and three in the afternoon, but she hadn't seen them lately. It developed that her kids had been shooting their sling-shots back there.

'Come back another day, *patrón*,' she said.

Now Don Alfredo sat gloomily on his bed in Mateo's lodge, con-sidering once again how to salvage Chelem-ha from the conflict, hunger, overpopulation, ignorance, deforestation, demoralization, persecution, and every other human, historic, and natural tragedy you could name, all of which surrounded the single mountain ridge, closing in fast. He began reasonably enough: 'It's prohibited to burn the forest for *milpa* without a licence from the departmental government. To get a licence, you need the owner's permission in writing. But nobody from the government ever comes up here to check. The people burn without a licence. If each family has ten or fifteen children, half may die, but still they're left with a lot of mouths to feed. This is the real problem here, but nothing's ever done about it. I can't protect Chelem-ha by myself . . .'

Mateo's son brought us the family's cheap red radio, which picked up Voice of America on medium wave. The announcer droned on incongruously, 'The United States State Department announced today . . .' something followed about the Communist threat. Nicar-aguan expansionism. Cuban terrorism. Soviet intervention . . .

In a few days' time I'll be back in the United States, I thought, where the gringo gods speak the private language of Cold War abstrac-tions. If only Central America were really so simple.

6

'Come back another day, *patrón*,' the woman inside the hut wall had said, but the next day broke badly – a fog so thick you could only see as far as the first row of maize surrounding Mateo's hut. The second row was already ghost maize. Out of that ghost *milpa* began to materialize misty men of maize, in a procession of ones and twos that went on for several hours before we left. They were coming to bring Don Alfredo their *cédulas* – their identification papers. They entered Mateo's lodge with humility, and Don Alfredo opened their documents and inspected each one. He worked from a list he'd brought with him, and that he was still compiling, of all the Indians presently living on Chelem-ha. When the list is completed, a contract will be written, and Chelem-ha's future will have been decided: the area will be deeded over to the Indians, who will make equal monthly payments out of their wages. They earn 100 quetzals a month – about $3 a day, the national minimum. It is probably a just solution, in a country never notable for justice. Whether it can halt the destruction of the Quetzal's cloud forest habitat is doubtful, but Don Alfredo will have to wait to find out.

We turned and rode slowly back through the canyon *milpas*, then uphill towards the pass. Don Alfredo was lost in thought, and so was I. Emilio trailed behind, stopping discreetly now and then to exchange words and glances with the Indian girls passing along the way. One was more lovely than the next, raising their round tawny faces above their embroidered *huipiles* and rainbow skirts. The beaded *puys* round their heads jangled with the rhythm of their movements. He would not remain a bachelor long. Little sun penetrated the dense green of the canyon pines. The people were on their way back from the *cafetal* to their *milpas*. The women strode with their proud and enchanting erect posture, balancing baskets, jugs, and cloth bundles on their heads. Their voices rose and fell in greeting to *El Patrón*. Don Alfredo greeted them in turn with a wistful and stoic grace.

At length I broke our silence with a question. 'Are you satisfied that it will come out this way, that the people will buy Chelem-ha?'

'Yes, I suppose so,' he said. 'Whatever's in *milpa* is already lost, so why not sell to them? I've lost hope that anyone will ever help me save the mountain as a wildlife refuge. The only chance now is to try to make an agreement with the people that they can work what they've got in *milpa* now, but leave alone what's left in forest. Mateo will help me. He understands.'

'You admire the Indians, don't you, Don Alfredo?'

He pushed his eyeglasses back up his nose again, and thought for a moment. 'Ah, yes,' he said finally. 'You can't be a tyrant with the people. There are some landowners who try to order everyone around and keep things strict, as in their father's day. That's not the way. It makes conflict, then the people rebel.'

And then we reached the pass, and the light dimmed. The highlands were filling up with the ominous colours of another rain. I took a last look back at the canyon, the smooth *milpas*, and Chelem-ha, and thought I could pick out several vultures, motionless dots in the brooding grey sky. But maybe that was only the future. The heartland of Central America blackened rapidly under the shadow of the coming storm.

Selected Bibliography

Books

Alvarado, Pedro de, two letters to Hernando Cortés reprinted in Patricia de Fuentes, *The Conquistadors*, Willows, Calif., Orion Press, 1963.

Amnesty International, *Guatemala: A Government Program of Political Murder* (Report), 1981.

—— *Guatemala: Massive Extrajudicial Execution in Rural Areas Under the Government of General Efraín Ríos Montt* (Special briefing), July 1982.

The Annals of the Cakchiquels, and Title of the Lords of Totonicapan, Norman, University of Oklahoma Press, 1953.

Asturias, Miguel Ángel, *Leyendas de Guatemala*, Buenos Aires, Editorial Losada, SA, 1957.

Coe, Michael D., *The Maya*, rev. ed., London, Thames & Hudson, 1980.

Dary Rivera, Mario, *Discursos universitarios*, Guatemala, Editorial Universitaria, 1982.

Davis, L. Irby, *A Field Guide to the Birds of Mexico and Central America*, Austin, University of Texas Press, 1972.

Davis, Nigel, *The Ancient Kingdoms of Mexico*, New York, Penguin Books, 1982.

Digby, Adrian, *Maya Jades*, London, The Trustees of the British Museum, 1972.

Hanke, Lewis, *The Spanish Struggle for Justice in the Conquest of America*, Boston, Little, Brown, 1949, 1965.

Hargreaves, Dorothy, and Hargreaves, Bob, *Tropical Blossoms of the Caribbean*, self-published, 1960.

—— *Tropical Trees*, self-published, 1965.

Janzen, Daniel H., ed., *Costa Rican Natural History*, Chicago, University of Chicago Press, 1983.

Kelsey, Vara, Osborne, Lilly, and Handel, Larry, *Four Keys to Guatemala*, New York, Funk & Wagnalls, 1978.

Kirkpatrick, F. A., *The Spanish Conquistadors*, London, Adam &. Charles Black, 1934.

Land, Hugh C., *Birds of Guatemala*, Wynnewood, Pa., Livingston Publishing, 1970.

Las Casas, Bartolomé de, *Tears of the Indians*, 1656. Translated by John Phillips, reprinted New York, Oriole Chapbooks, 1972.

Long, Trevor, and Bell, Elizabeth, *La Antigua Guatemala: a Portrait of a Colonial Capital*, La Antigua, Guatemala, Editorial La Galería, 1979.

Meneses, Carlos, *Miguel Ángel Asturias*, Barcelona, Ediciones Jucar, 1975.

Morley, Sylvanus, *The Ancient Maya*, Stanford, Calif., Stanford University Press, 1956.

Nations, James D., and Komer, Daniel I., *Conservation in Guatemala*, Austin, Texas, Report of the Center for Human Ecology, 1984.

Nicholson, Irene, *Mexican and Central American Mythology*, London, Paul Hamlyn, 1967.

North American Congress on Latin America, *Guatemala*, New York, 1981.

Paz, Octavio, *Conjunctions and Disjunctions*, New York, Grove Press, 1969.

Perry, Frances, and Hay, Roy, *A Field Guide to Tropical and Subtropical Plants*, New York, Van Nostrand Reinhold, 1982.

Peterson, Roger Tory, and Chalif, Edward L., *A Field Guide to Mexican Birds*, Boston, Houghton Mifflin, 1973.

The Popol Vuh, English version by Delia Goetz and Sylvanus G. Morley from the translation of Adrian Recintos, 12th ed., Norman, University of Oklahoma Press, 1983.

Rodman, Selden, *The Guatemala Traveler*, Meredith Press, 1967.

Rosenthal, Mario, *Guatemala: Emergent Latin-American Democracy*, Twayne Publishers, 1962.

Salvin, Osbert, *Biologia centrali americana*, London, A. P. Maudsley, 1902.

Silvert, K. H., *Guatemala: A Study in Government*, New Orleans, Tulane University Middle American Research Institute.

Skutch, Alexander F., *The Imperative Call*, Gainesville, University of Florida Press, 1979.

Smithe, Frank B., and Trimm, H. Wayne, *The Birds of Tikal*, Garden City, N.Y., The Natural History Press, 1966.

Stephen, David, and Wearne, Phillip, *Central America's Indians*, London, Minority Rights Group, 1984 (Report no. 62).

Stephens, John L., *Incidents of Travel in Central America, Chiapas, and Yucatán*, New York, Harper & Bros., 1847. Reprinted New York, Dover Publications, 1969.

Thompson, Edward Herbert, *People of the Serpent*, New York, Capricorn Books, 1926.

Thompson, J. Eric S., *The Rise and Fall of Maya Civilization*, 6th ed., Norman, University of Oklahoma Press, 1977.

Turner, Anne Warren, *Vultures*, New York, David McKay, 1973.

Wolf, Eric R., *Sons of the Shaking Earth*, Chicago, University of Chicago Press, 1959.

Young, John Park, *Central American Currency and Finance*, Princeton, N.J., Princeton University Press, 1925.

Zweig, Stephan, *The Tide of Fortune*, New York, Viking Press, 1940.

Periodicals

Aveni, Anthony F., 'Venus and the Maya', *American Scientist*, 67, May–June 1979, pp. 274–85.

La Bastille, Anne, 'The Quetzal', *National Geographic*, January 1969, pp. 141–50.

La Bastille, Anne, Allen, D. G., and Durrell, L. W., Behavior and Feather Structure of the Quetzal', *The Auk*, 89, April 1972, pp. 339–48.

MacNeish, Richard S., 'The Origins of New World Civilization', *Scientific American*, 211, no. 5, November 1964, pp. 29–37.

Peterson, Roger Tory, 'Vulture Vigils on Four Continents', *Audubon*, November–December 1968, pp. 82–91.

Salvin, Osbert, 'Quesal-shooting in Vera Paz', *The Ibis*, 1861, pp. 138–49.

Skutch, Alexander F., 'The Life History of the Quetzal', *The Condor*, 46, no. 5, September–October 1944, pp. 213–35.

Index